press start to translate

This is what happens when you let a computer translate a video game?

Legends of Localization

by Clyde Mandelin

First Edition - November 2017

Legends of Localization is an examination of the cultural influences and differences that affect Japanese and English games. *press start to translate* is an unofficial, educational, history-focused exploration of classic games. Certain portions of this book's content (and the localization information on many other games) are freely available online at: *http://legendsoflocalization.com*

press start to translate is designed, published, sold, and distributed by Fangamer, and printed by the hardworking folks at Bang Printing in Palmdale, CA.

ISBN: 978-1-945908-86-6

Team

Clyde Mandelin
Concept, Writing, Research

Tony Kuchar
Design, Layout

Heidi Mandelin
Producer, Research, Editing

Dan Moore
Editing, Blurb Writing

Ryan Novak
Editing

Reid Young
Assistant Producer, Boss Man, Candy Enthusiast, Black Licorice Hater

O.G. San
Model, Mascot, Spirit Animal

Special Thanks

byuu

Kari Fry

Nina Matsumoto

Revenant

insightblue

Lenalia?

All the PK Hack dudes

MightyYT

Kent Hansen

Expert Commercial Software

Genecyst East Software

NeoPaula & diospadre

spartonberry

Magma Sushi Lounge

Shota Sushi and Grill

Guilin Chinese Restaurant

Kuze

Researchers and engineers at Google

GIHEIYA

Everyone who helped
on Twitter & Twitch

Contents

Introduction

Technology has reshaped our world in less than a single lifetime. To quote a classic 1980s movie: "Everything is different, but the same. Things are more moderner than before. Bigger, and yet smaller... It's computers..." This comment rings truer now than ever before.

We rely on computers and automated tools for nearly everything today. We depend on software to catch our spelling mistakes, to give us driving directions, and to find our soulmates. We count on automated tools for medical diagnoses, weather predictions, and traffic control. And, of course, we rely on computers for global communication, even when language barriers stand in the way.

The idea of using machines to translate between languages is surprisingly old. The earliest concepts date back to the 1600s, but it wasn't until the 1950s that research into computer-based machine translation began.

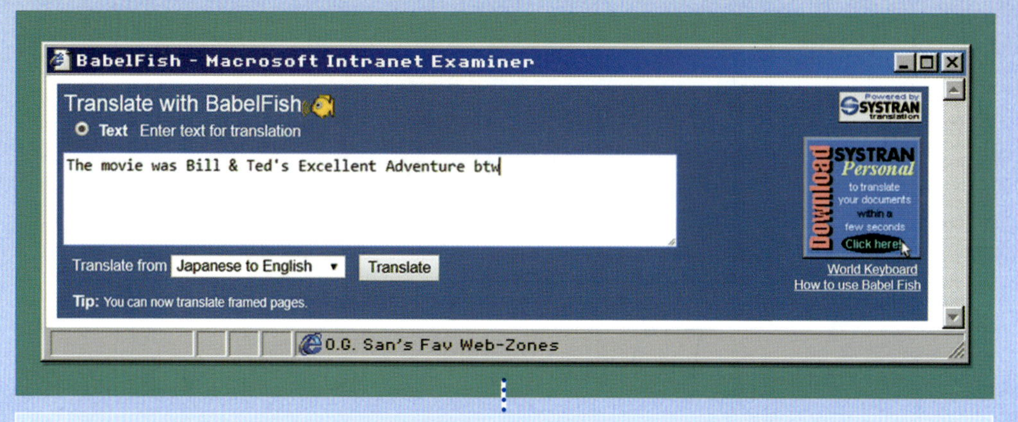

AltaVista's BabelFish site, which debuted in 1997, introduced millions of people to the idea of machine translation.

Today, machine translation technology is accessible to anyone with a computer or phone.

Despite its wide availability and ease of use, machine translation has a reputation for being unreliable. This stigma began with online translation tools like AltaVista's BabelFish in the late 1990s and remains strong today in spite of newer tools and smartphone apps. Consequently, this reputation has indirectly highlighted the value of professional, human-based translation.

Of course, machine translation technology continues to improve at a rapid pace. At the same time, companies rely on machine translation more and more as whole industries expand across the world. This is especially evident in the ever-growing video game industry – machine translation offers a speedy, low-cost way for game publishers to reach new markets and increase profits. This convenience comes at a price, of course, but how bad can it really be? What could possibly go wrong?

In this book, we'll take a detailed look at what can happen when a machine translates a full-length video game from Japanese into English. Along the way, we'll also:

- See how translation technology suddenly evolved during this project
- Compare hundreds of translated lines with their original, intended meanings
- Explore what machines get right, what they get wrong, and why
- See how machines handle common obstacles that all translators face

Don't let all of this fancy talk fool you, though – this is meant to be a breezy, laid-back look at an incredibly deep topic. By the end, we'll have learned a lot about machine translation, human translation, the Japanese language, game reprogramming, and eagles. It's my hope that you'll find this book entertaining, insightful, and informative. And who knows, maybe we'll have some good laughs too!

As this example from *Alice in the Heart ~Wonderful Wonder World~* demonstrates, machine translations can be hard to differentiate from poor human translations. By the end of this book, you'll be able to spot some of the signs of Japanese-to-English machine translation.

Solo programmers and indie game studios regularly rely on machines for their translation needs. This example from *Slash Hero* illustrates what can happen if a developer is unfamiliar with the pitfalls of the machine translation process.

Project Background

My name is Clyde Mandelin, although I'm better known online as "Tomato" or simply "Mato". I've been a professional Japanese-to-English translator for 15 years now, with a focus on movies, games, anime, and other entertainment. I also use my computer science background to tinker with unofficial hobby translation projects – I basically take old, untranslated video games and translate them into English.

More recently, I combined my interests to create *Legends of Localization*, a series of articles and books about how video games change during translation. I also use real examples from actual games to illustrate how the translation process works. By focusing on real-life examples, I'm able to explain complex theoretical concepts to non-translators. It's a fun and rewarding process that helps me grow as a translator too.

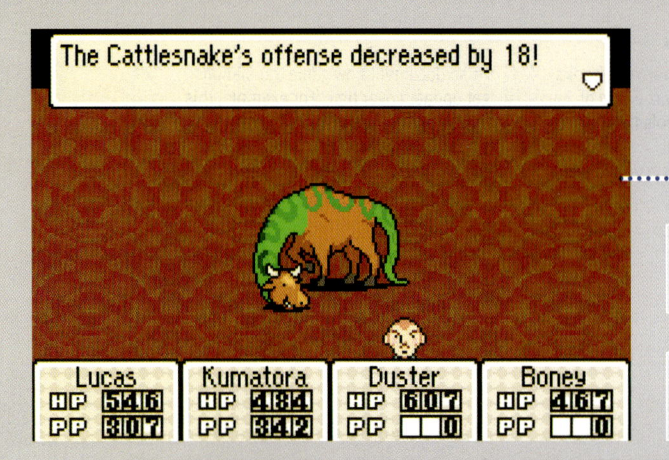

One of my old hobby projects involved translating *MOTHER 3*, the sequel to *EarthBound*, into English for fellow fans.

Years upon years of my professional life have flowed through my keyboard and into all these little boxes I store in my closet.

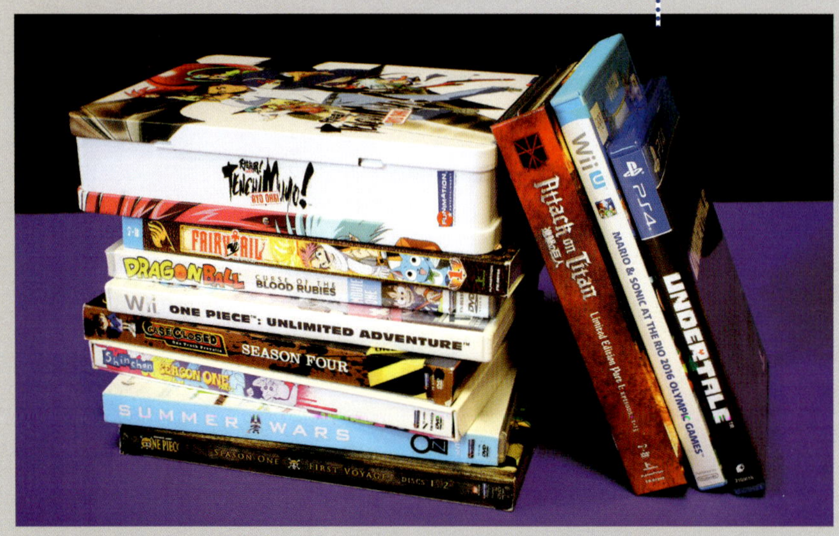

I recently branched out into writing books about video game translation. Some of them are even used in schools now!

I also document how game translations get updated over time. For example, this line from *The Legend of Zelda* was revised 30 years after its original release!

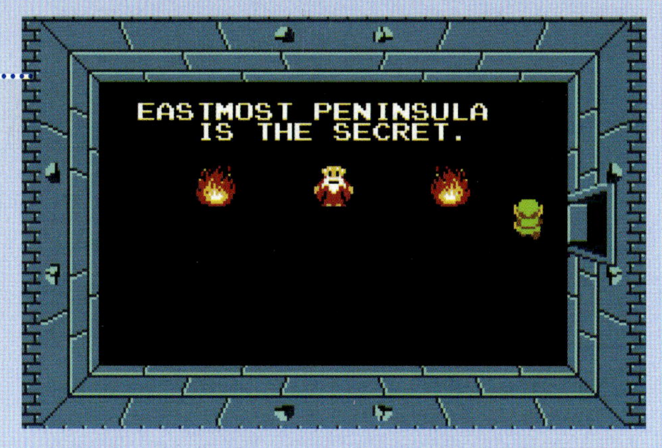

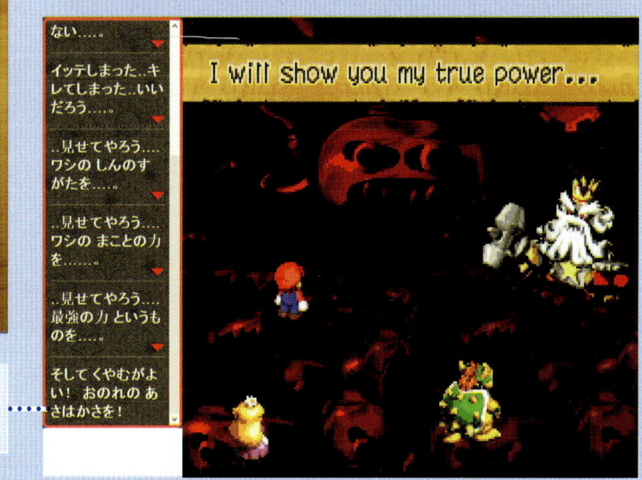

I sometimes program nifty things like this tool that displays a game's Japanese script as the English game is being played.

Original Plan

After I started *Legends of Localization*, I began to daydream about a crazy, farfetched project. I wanted to design a special software tool that could remove all of the text from a video game and replace it with entirely new text. With this tool, I could insert brand-new translations into well-known games. This would be helpful in several ways:

- I could easily compare different translation styles in action, side by side, without resorting to theoretical discussion. How exactly does a literal translation compare to a liberal translation, for example?
- I could show off the problems that are unique to video game translation, such as memory limitations and glitchy effects caused by a single typo.
- I could examine how external factors like team sizes and deadlines affect the quality of a translation.

In a sense, translators are explainers at heart, so it's not surprising that we like to explain the translation process too. In this notorious example from *Rhapsody: A Musical Adventure*, a translation staff member broke the fourth wall to make a point about direct translations.

For a long time, my translation tool idea was nothing more than a "wouldn't it be cool" fantasy. Then, one day in late 2016, I suddenly decided to make it happen. I spent the next few weeks designing software and writing code. Before long, I had a prototype of my tool hooked up to a classic Japanese video game known as *Final Fantasy IV*. There was one problem, though: I needed a quick and easy way to test the tool, but crafting a new translation by hand would take weeks. That's when I turned my attention to machine translation.

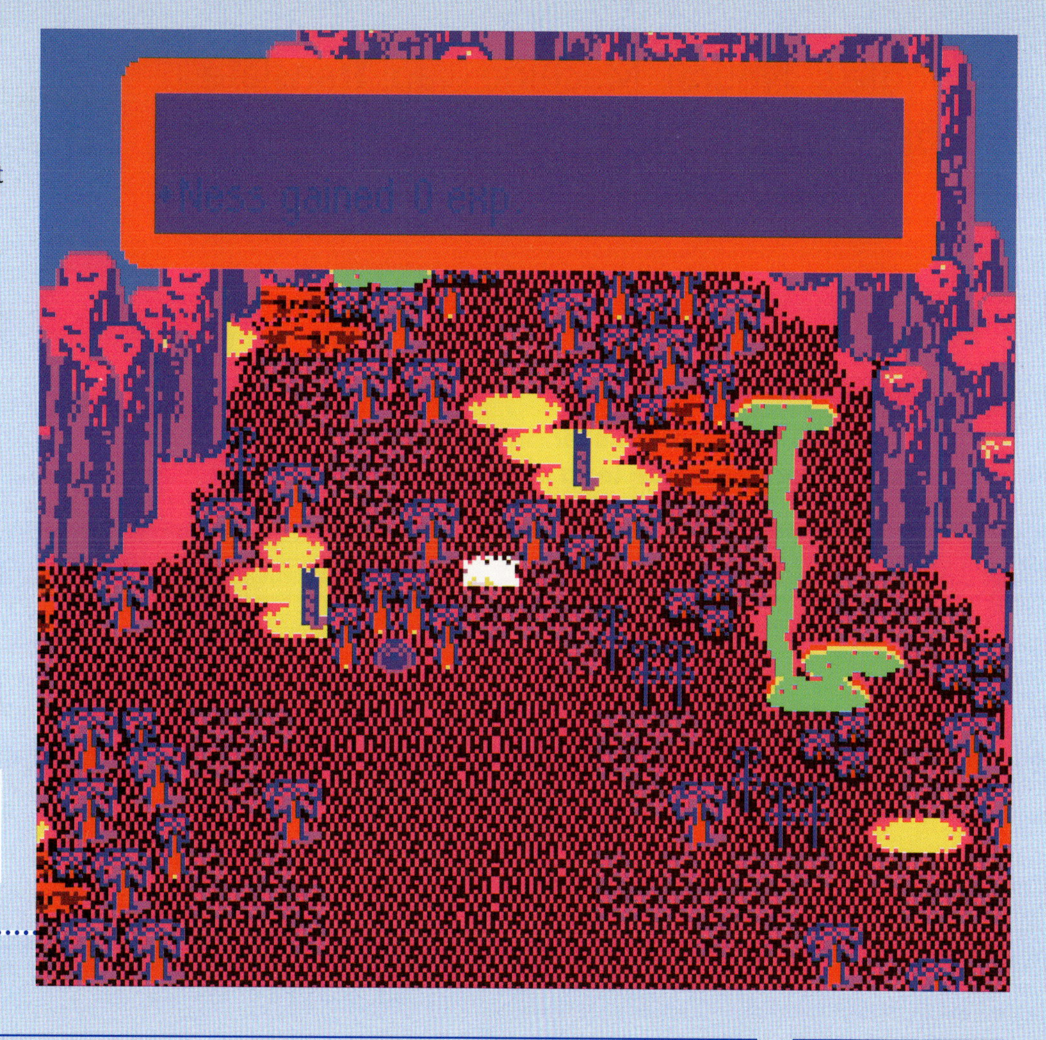

Every video game is a complex program – if even one little thing gets out of place, the entire game can fall apart. Trying not to break anything is an important part of the game translation process.

Google's Neural Machine can translate nearly as well as a human

LAST UPDATED ON OCTOBER 5TH, 2016 AT 7:20 PM BY ALEXANDRU MICU

A new translation system unveiled by Google, the Neural Machine Translation (GNMT) framework comes close to human translators in it's proficiency.

INTERNET

Google Translate just got a lot smarter

The search giant says it's made a "leap" in giving you more natural translations. It's just one step in a big push in machine learning that Google detailed Tuesday.

BY RICHARD NIEVA / NOVEMBER 15, 2016 12:54 PM PST

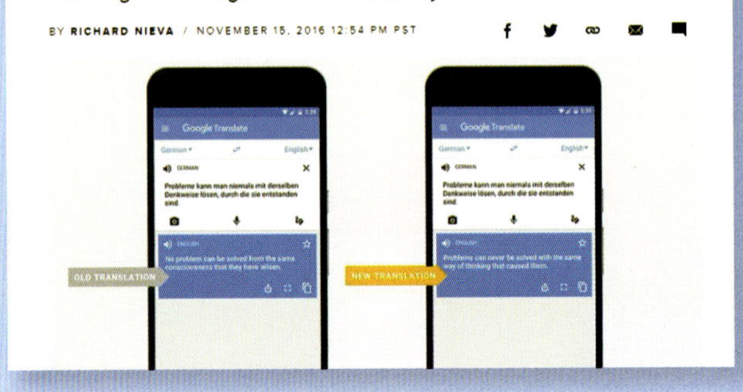

Google Translate: 'This landmark update is our biggest single leap in 10 years'

Google Translate should make far fewer errors thanks to Google's new neural machine-translation technology, which is now rolling out to eight languages.

By Liam Tung | November 17, 2016 -- 13:42 GMT (05:42 PST) | Topic: Artificial Intelligence

Translation quality
6
5
4
3
2
1
0
perfect translation
human
neural (GNMT)
phrase-based (PBMT)
English > Spanish
English > French
English > Chinese
Spanish > English
French > English
Chinese > English
Translation model

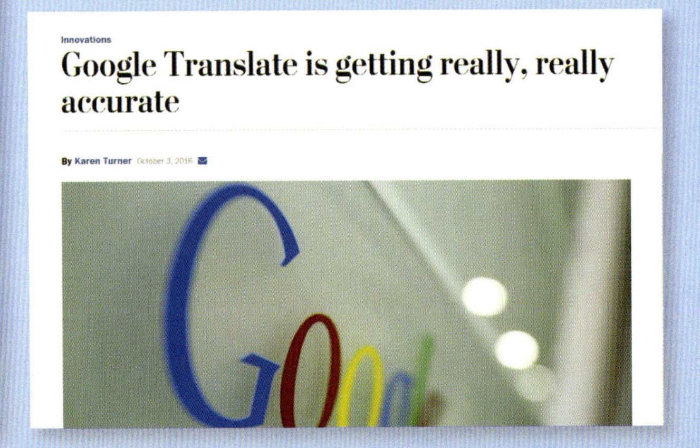

Innovations

Google Translate is getting really, really accurate

By Karen Turner October 3, 2016

Getting Funky

In December 2016, I ran *Final Fantasy IV*'s Japanese text through an online tool known as Google Translate. In the blink of an eye, hundreds of pages of Japanese writing transformed into English. I then used my custom software to insert this English text back into *Final Fantasy IV*.

My first test was a near-perfect success – the game displayed the new text exactly as Google Translate wrote it. The machine translation was (unsurprisingly) pure nonsense, so I gave this version of *Final Fantasy IV* a suitable new name: *Funky Fantasy IV*.

Several days later, I ran *Final Fantasy IV*'s text through Google Translate again as part of another test. To my surprise, this second translation was incredibly different from the first one. I didn't know it at the time, but Google had upgraded its machine translator to use cutting-edge neural network artificial intelligence. By sheer luck, I had taken before-and-after snapshots of this technology in action – on a large scale, no less!

I immediately began to document both translations, mostly out of personal curiosity, but also as a fun *Legends of Localization* project. As I played through both translations, I took screenshots of interesting lines and streamed the entire experience online for others to enjoy.

I also shared a number of screenshots on a *Funky Fantasy IV* web page. The response was so strong that I soon found myself digging deeper into the project – deep enough to write a book about it.

In the pages ahead, we'll briefly look at *Funky Fantasy IV*'s development process. We'll survey the first machine translation – which we'll call "Translation #1" – followed by a more thorough look at Translation #2 from start to finish.

After that, we'll compare both translations to see how things changed for the better or worse. We'll also evaluate a machine's reading comprehension ability and its potential for improvement.

Finally, we'll check out some of the real-life surprises that ensued after news of *Funky Fantasy IV* got out. For an unplanned, spur-of-the-moment project, there's a whole lot to talk about!

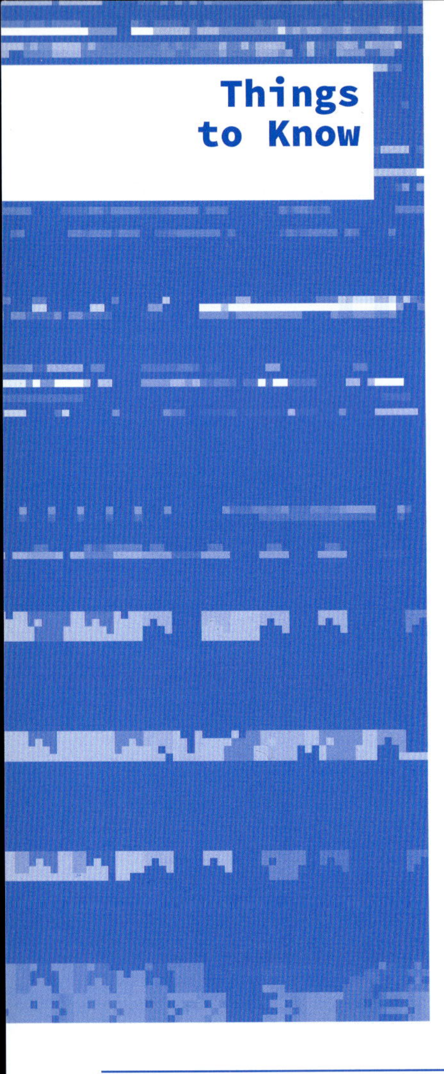

Things to Know

Before we get too deep into all the machine translation fun, here are a few tips on how to make the most of the pages ahead.

Three Ways to Read This Book

Because this is such an unusual book about an unusual topic, we designed it so that there are three different ways to read through it:

1. For laughs and entertainment, you can skip everything and focus on the screenshots
2. For curiosity's sake, you can consult the proper translations listed beneath each screenshot
3. For further details and insight, you can read through my accompanying commentary

Newcomers Welcome

This book assumes no prior knowledge of the Japanese language or *Final Fantasy IV*. When needed, I provide extra details to explain what's going on.

1	バカ	Katakana	*baka* ("idiot")
2	ばか	Hiragana	*baka* ("idiot")
3	馬鹿	Kanji	*baka* ("idiot")

Despite how different they look, all three of these are the same exact word. As we can see, the *kanji* version is much more complex than the other two.

Japanese Writing Systems

Briefly put, the Japanese language uses three writing systems: *katakana*, *hiragana*, and *kanji*. Ordinary Japanese writing uses a balance of all three, but early games like *Final Fantasy IV* only use the first two. For humans, this imbalance makes text a chore to read. For machines, it causes all sorts of problems that we'll examine in the coming chapters. For now, just keep the words *katakana* and *hiragana* in mind.

Weird Vowels

From time to time, you might see Japanese vowels in this book with weird lines above them, like *ā* and *ō*. Basically, these vowels are held a little longer than usual when pronounced out loud.

Are you guys giants or something? A

B
Story
Titus the tailor told ten tall tales to Titania the titmouse. Oh, her bridal bower becomes a burial bier of bitter bereavement!
C
Mato Says
If you are reading this then you are exceptionally smart and attractive. Also, the Japanese *kanji* for "poop" is 糞.
D
Reader Challenge
Quick! What is the longest English word you can think of? Is it "shoelace"?

Reading Guide

In the chapters ahead, you'll see all kinds of boxes and sidebars that accompany each set of game images. Don't be scared and confused – this handy guide will explain how everything works.

A. Intended Meanings: The captions beneath each game screenshot show what a proper translation of the Japanese text would be. Note that these are my own professional translations and are not based on any previous translations of *Final Fantasy IV*.

B. Story Box: *Final Fantasy IV*'s story goes all over the place, and the machine translations make the story even harder to follow along. If you're unfamiliar with *Final Fantasy IV*, these story snippets will help you make sense of it all.

C. Mato Says: Every so often, I'll share personal and professional insight in these tomato-themed boxes.

D. Challenge Box: You can learn a lot by reading about translation, but you'll discover even more by wrestling with translation yourself. Take on these unique challenges... if you dare.

E. Page Number: This is the current page number, written out in Japanese *hiragana* and then machine-translated into English using the same system as Translation #2. See if you can make sense of any of it!

Getting Technical

Development

Going from the simple "Hey, let's machine-translate an entire game" stage to an actual, fully functional, machine-translated game is a big jump. So before we dive into *Funky Fantasy IV*'s text, let's take a quick look at the technical side of things and see how everything works under the hood.

Why Final Fantasy IV?

There are probably tens of thousands of games to play around with, so why pick *Final Fantasy IV* for this project? What makes it special?

First, *Final Fantasy IV* was released in the early 1990s, when console games were just beginning to level up technologically. Earlier console games are messier to modify due to technical limitations, and later games are much more complicated. In short, *Final Fantasy IV* is a very simple, yet robust game to work with.

Final Fantasy IV also marked the start of a new era of storytelling in Japanese video games. In contrast to earlier console games, *Final Fantasy IV* boasted a strong, brisk plot filled with colorful characters, surprise twists, and imaginative writing. The game's story is often likened to *Star Wars*, but I'd say it's more like a mix of *Star Wars*, soap opera, and ancient Greek epic, with a dash of Shakespearean intrigue for added flavor. In this way, *Final Fantasy IV* is a true classic and an important piece of video gaming history.

Even more unique about *Final Fantasy IV*, though, is how often it's been retranslated. It's not uncommon for classic works of literature to receive more than one English translation – examples include *Don Quixote* (12+ translations) and *The Tale of Genji* (6+ translations). But those books were originally published in the early 1600s and the early 1000s. *Final Fantasy IV* debuted in 1991, yet it now has five separate English translations – and even more if you include minor variations and unofficial fan-made translations. By some quirk of fate, *Final Fantasy IV* and translation just go hand-in-hand.

Lastly, I'm a longtime fan of *Final Fantasy IV* and know the game inside and out. I've also been studying the game's multiple translations for years now as part of another *Legends of Localization* project. Given all of this, *Final Fantasy IV* seemed like the most logical choice for my translation experiment.

Final Fantasy IV keeps getting so many remakes, re-releases, and re-translations that it's hard to keep track of them all!

Inside the Game

On the outside, video games are fun and exciting to look at. On the inside, they're complex, tangled webs of 1s and 0s. Old console games are especially difficult to untangle, so my idea of replacing *Final Fantasy IV*'s text presented a challenge. Ordinary software like Microsoft Office and Notepad isn't designed to modify games, which meant I had to dive into the 1s and 0s myself and create my own tools. Specialized technology lies at the very heart of the *Funky Fantasy IV* project, so let's take a brief, technical look at its development.

Going for Full Automation

My original plan was to create a tool that could extract text from a game and insert new text in its place, all without human intervention. This meant that things like text formatting and text abbreviation needed to be handled automatically. This, combined with a machine translation system, would make the entire translation process 100% automated. I almost reached that goal, but a few unexpected snags dropped that number down to about 95%.

Development Recipe

Funky Fantasy IV's development was broadly divided into seven different stages:

1. Deciphering *Final Fantasy IV*'s unique text format
2. Reverse engineering the game's multiple text storage systems
3. Locating all of the text data
4. Extracting the Japanese text into standard computer text files
5. Running the text through Google Translate
6. Inserting the new English text back into the game
7. Reprogramming the game to load the English text properly

This little microchip is the CPU of the Super NES game console. I spent a large portion of the project studying how this old doohickey worked.

After about three weeks of work, I had a fully machine-translated version of *Final Fantasy IV* up and running. Most of that time was spent staring at numbers and computer code. The actual translation part only took a few seconds, thanks to the magic of machine translation.

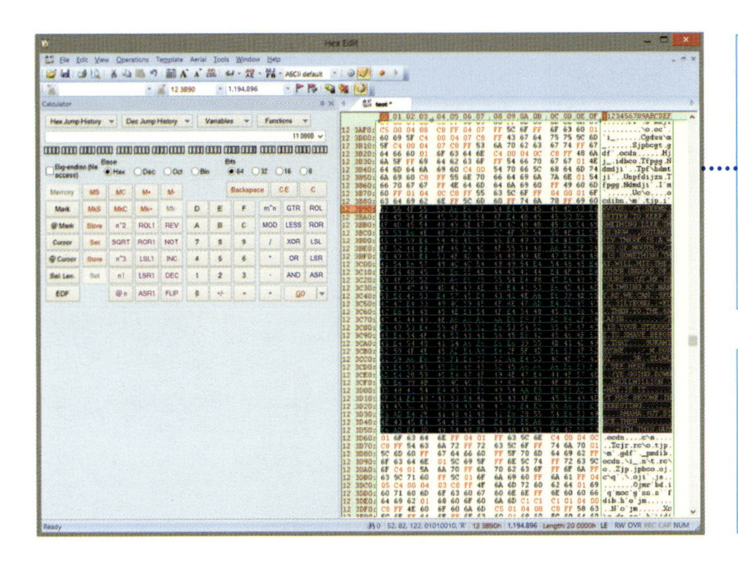

Old console games are basically enormous blobs of numbers all in one file, so I used my trusty hex editor to sift through all of *Final Fantasy IV*'s data. In simple terms, it was like looking at the game with an X-ray machine.

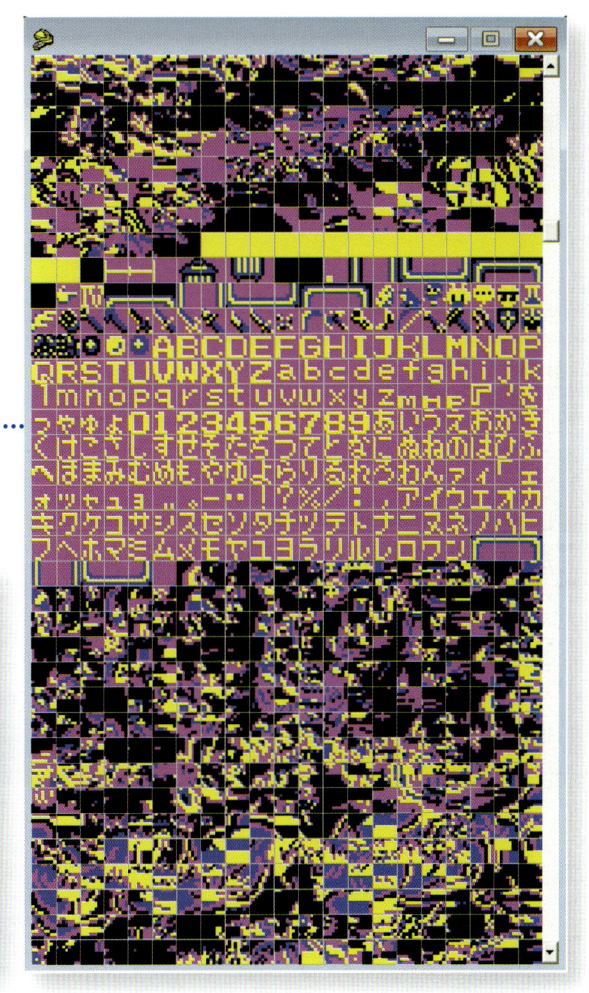

I used a tile editor to locate the game's font data. This data helped me decipher *Final Fantasy IV*'s proprietary text format, similar to how a decoder ring works.

Locating text data among all the other game data was tough, so I used another tool to manually scan for Japanese text. This was a lot like using an MRI machine and radioactive dye to highlight what I was looking for.

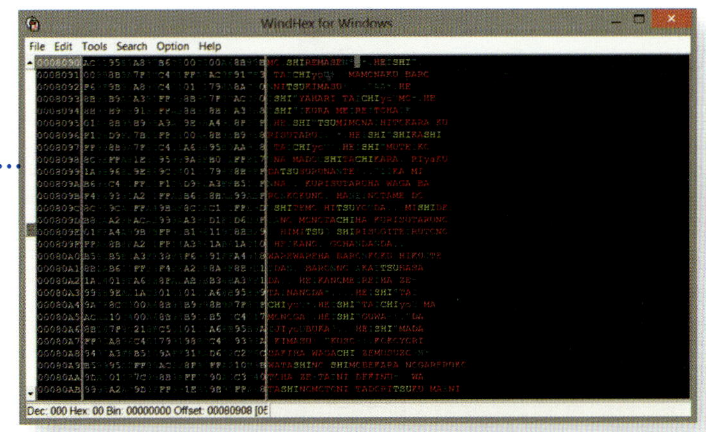

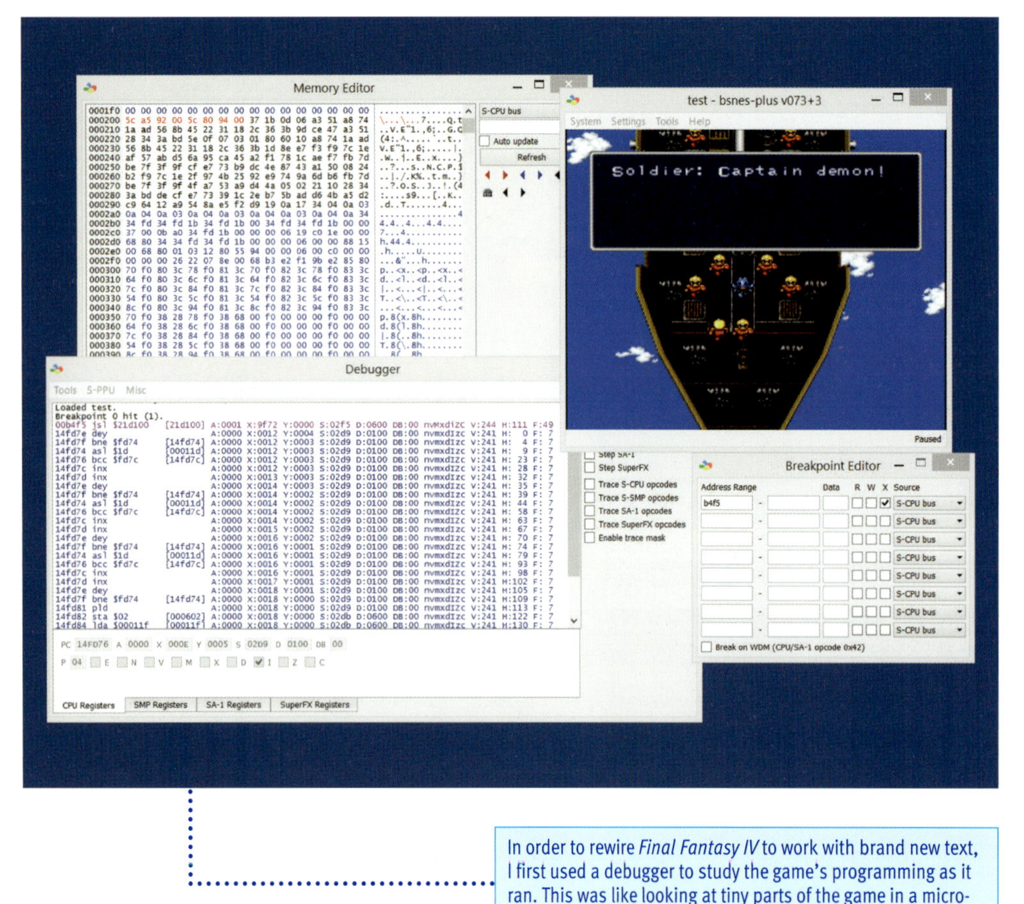

In order to rewire *Final Fantasy IV* to work with brand new text, I first used a debugger to study the game's programming as it ran. This was like looking at tiny parts of the game in a microscope to see what was going on at the lowest level.

Next, I learned Super NES programming so I could redesign the game to work with a brand new script. English scripts are usually two to three times larger than their Japanese counterparts, so I had to find a way to cram lots of extra text into the game.

```
 89 bcs iffb
 90 LDA $268000,x
 91 jmp byebye2b
 92
 93 iffb:
 94 rep #$20
 95
 96 txa
 97 sbc #$8000
 98 tax
 99 lda #$0000
100 sep #$20
101 LDA $278000,x
102
103 byebye2b:
104 plx
105 RTL
106
107
108 //===========================================
109 // TEMP JUMP-OFF AREA NEAR END OF FIRST BANK
110 //===========================================
111 org $1ffe8
112 preloadwindow:
113 jsl loadwindow
114 rts
115
116
117 //===========================================
118 // RELOCATING THE INDIVIDUAL STATUS WINDOW DATA
119 //===========================================
120
121 // we're gonna load the new menu data, which we
122 org $1abbc
123 ldy #$0100
124
125 org $18307
126 jsr preloadwindow
127 nop
```

I created my own custom tools to extract the Japanese text, reinsert the translated text, and modify other text as needed. This was like doing surgery to remove a defective heart and then replacing it with an artificial heart.

The actual translation process was the easiest: I simply fed the machine translator the Japanese text files, clicked a menu option, and saved the translation to a new text file.

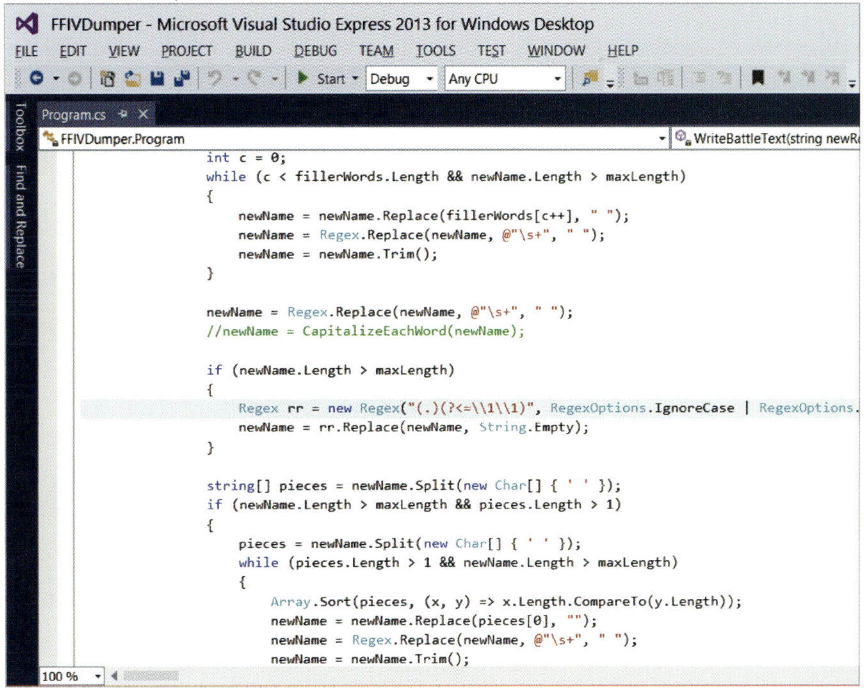

FFIVDumper - Microsoft Visual Studio Express 2013 for Windows Desktop

FILE EDIT VIEW PROJECT BUILD DEBUG TEAM TOOLS TEST WINDOW HELP

Start Debug Any CPU

Program.cs

FFIVDumper.Program — WriteBattleText(string newR

```
int c = 0;
while (c < fillerWords.Length && newName.Length > maxLength)
{
    newName = newName.Replace(fillerWords[c++], " ");
    newName = Regex.Replace(newName, @"\s+", " ");
    newName = newName.Trim();
}

newName = Regex.Replace(newName, @"\s+", " ");
//newName = CapitalizeEachWord(newName);

if (newName.Length > maxLength)
{
    Regex rr = new Regex("(.)(?<=\\1\\1)", RegexOptions.IgnoreCase | RegexOptions.
    newName = rr.Replace(newName, String.Empty);
}

string[] pieces = newName.Split(new Char[] { ' ' });
if (newName.Length > maxLength && pieces.Length > 1)
{
    pieces = newName.Split(new Char[] { ' ' });
    while (pieces.Length > 1 && newName.Length > maxLength)
    {
        Array.Sort(pieces, (x, y) => x.Length.CompareTo(y.Length));
        newName = newName.Replace(pieces[0], "");
        newName = Regex.Replace(newName, @"\s+", " ");
        newName = newName.Trim();
```

100 %

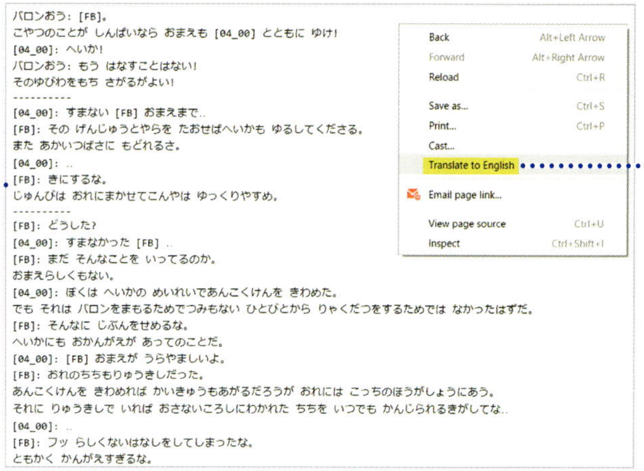

```
バロンおう: [FB]。
こやつのことが しんぱいなら おまえも [04_00] とともに ゆけ!
[04_00]: へいか!
バロンおう: もう はなすことはない!
そのゆびわをもち さがるがよい!
----------
[04_00]: すまない [FB] おまえまで..
[FB]: その げんじゅうとやらを たおせばへいかも ゆるしてくださる。
また あかいつばさに もどれるさ。
[04_00]: ..
[FB]: きにするな。
じゅんびは おれにまかせてこんやは ゆっくりやすめ。
----------
[FB]: どうした?
[04_00]: すまなかった [FB] ..
[FB]: まだ そんなことを いってるのか。
おまえらしくもない。
[04_00]: ぼくは へいかの めいれいであんこくけんを きわめた。
でも それは バロンをまもるためでつみもない ひとびとから りゃくだつをするためでは なかったはずだ。
[FB]: そんなに じぶんをせめるな。
へいかにも おかんがえが あってのことだ。
[04_00]: [FB] おまえが うらやましいよ。
[FB]: おれのちちもりゅうきしだった。
あんこくけんを きわめれば かいきゅうもあがるだろうが おれには こっちのほうがしょうにあう。
それに りゅうきしで いれば おさないころしにわかれた ちちを いつでも かんじられるきがしてな..
[04_00]: ..
[FB]: フッ らしくないはなしをしてしまったな。
ともかく かんがえすぎるな。
```

Back Alt+Left Arrow
Forward Alt+Right Arrow
Reload Ctrl+R
Save as... Ctrl+S
Print... Ctrl+P
Cast...
Translate to English
Email page link...
View page source Ctrl+U
Inspect Ctrl+Shift+I

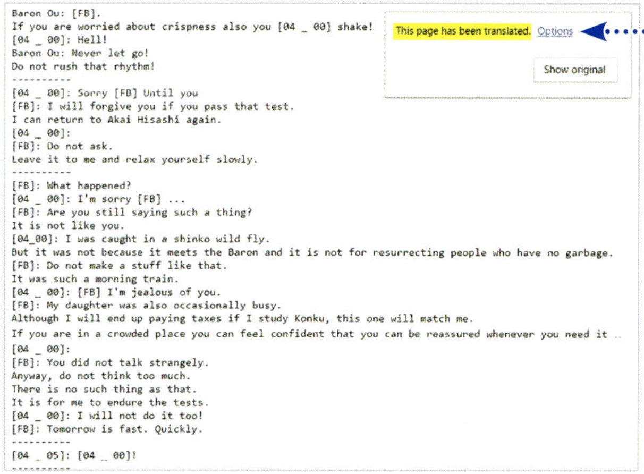

```
Baron Ou: [FB].
If you are worried about crispness also you [04 _ 00] shake!
[04 _ 00]: Hell!
Baron Ou: Never let go!
Do not rush that rhythm!
----------
[04 _ 00]: Sorry [FB] Until you
[FB]: I will forgive you if you pass that test.
I can return to Akai Hisashi again.
[04 _ 00]:
[FB]: Do not ask.
Leave it to me and relax yourself slowly.
----------
[FB]: What happened?
[04 _ 00]: I'm sorry [FB] ...
[FB]: Are you still saying such a thing?
It is not like you.
[04_00]: I was caught in a shinko wild fly.
But it was not because it meets the Baron and it is not for resurrecting people who have no garbage.
[FB]: Do not make a stuff like that.
It was such a morning train.
[04 _ 00]: [FB] I'm jealous of you.
[FB]: My daughter was also occasionally busy.
Although I will end up paying taxes if I study Konku, this one will match me.
If you are in a crowded place you can feel confident that you can be reassured whenever you need it ..
[04 _ 00]:
[FB]: You did not talk strangely.
Anyway, do not think too much.
There is no such thing as that.
It is for me to endure the tests.
[04 _ 00]: I will not do it too!
[FB]: Tomorrow is fast. Quickly.
----------
[04 _ 05]: [04 _ 00]!
----------
```

This page has been translated. Options

Show original

Trimming the Text

Not all of the text in *Final Fantasy IV* is character dialogue – there's plenty of secondary text that appears in menus, battles, and special events. Most of it is very short and intended to fit inside tiny text boxes. This produces a new problem: what do we do if our new, machine-translated text is too long?

The problem of squeezing long text into limited space has been around since the beginning of video game translation. For this project, I designed my software to automatically shorten secondary text in a number of ways:

- Cutting unneeded spaces and punctuation
- Dropping minor words like "the" and "a"
- Removing certain vowels and consonants

Different types of text followed different steps. The results weren't always optimal, but my shortening system worked well enough to make the game playable.

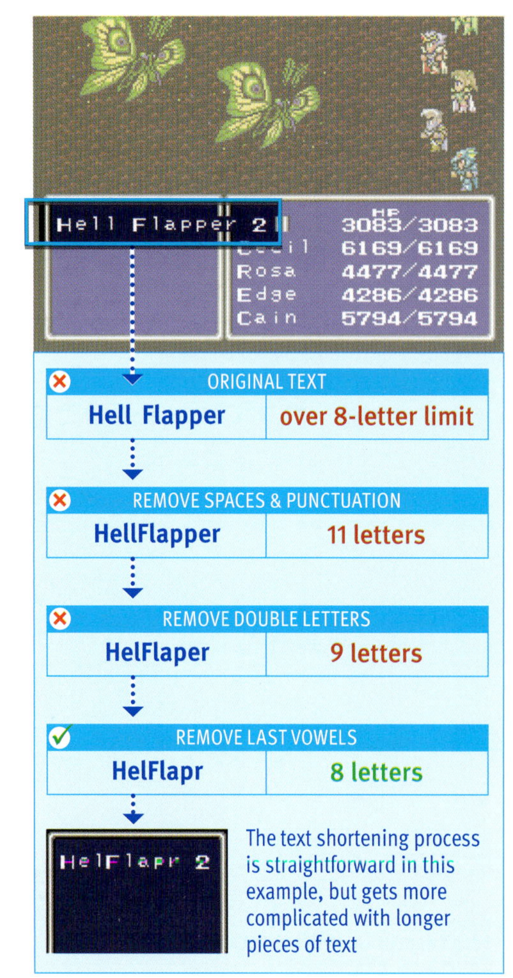

The text shortening process is straightforward in this example, but gets more complicated with longer pieces of text

(*Unshortened Text: "A monster left from the door!"*)

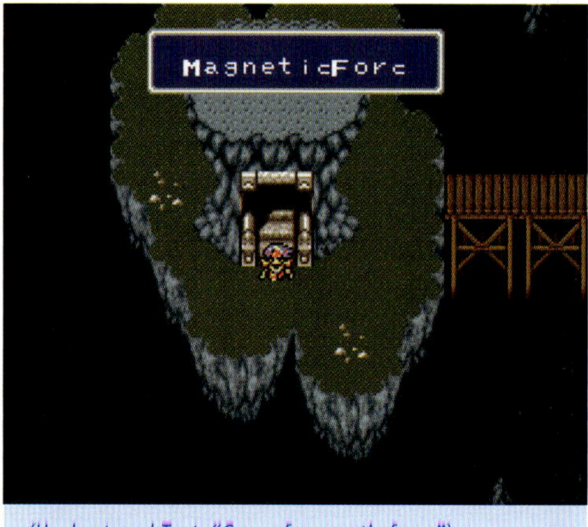

(*Unshortened Text: "Cave of magnetic force"*)

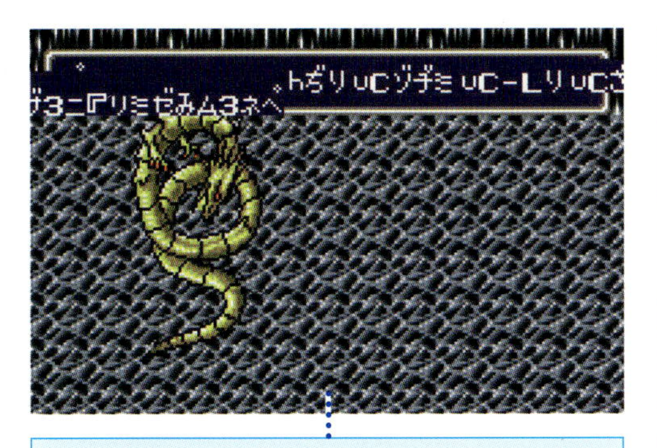

Reprogramming the game's text systems involved a lot of trial and error, but mostly error.

I quickly learned that machines can't be trusted to handle special name and number markers properly. I had to fix these issues manually.

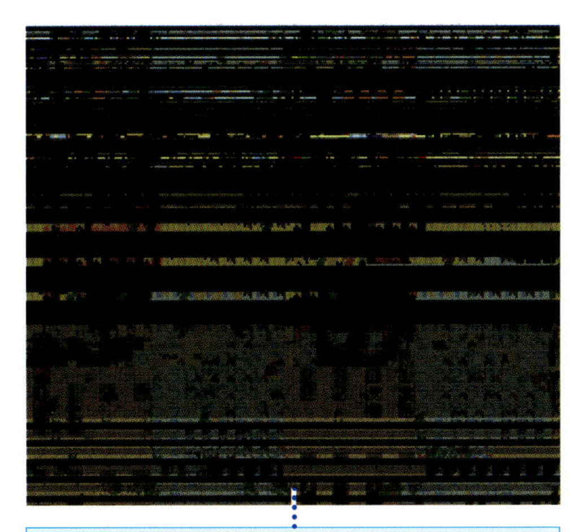

Oops, I accidentally disintegrated the moon.

I don't want to live on this planet anymore.

Goofs & Glitches

Writing software is a test of patience – bugs appear out of nowhere and programs rarely work properly on their first test run.

During the several weeks I spent working on *Funky Fantasy IV*, I caused all kinds of crashes, glitches, and disasters. Some of my bugs resulted in garbled text everywhere. Others exploded the game's graphics into a mangled mess. One time, I got the game to honk at me whenever the main character took a step. I fixed these issues, of course, but somehow they felt fitting for such a silly project.

Getting Down to Business

Most of my technical work for *Funky Fantasy IV* wrapped up in early December 2016. A few tiny issues remained, but I was eager to get on to the main attraction: inserting the new text and playing through a 100% machine-translated version of *Final Fantasy IV*. I didn't know what to expect, nor did I have any idea what was to come just a few days later!

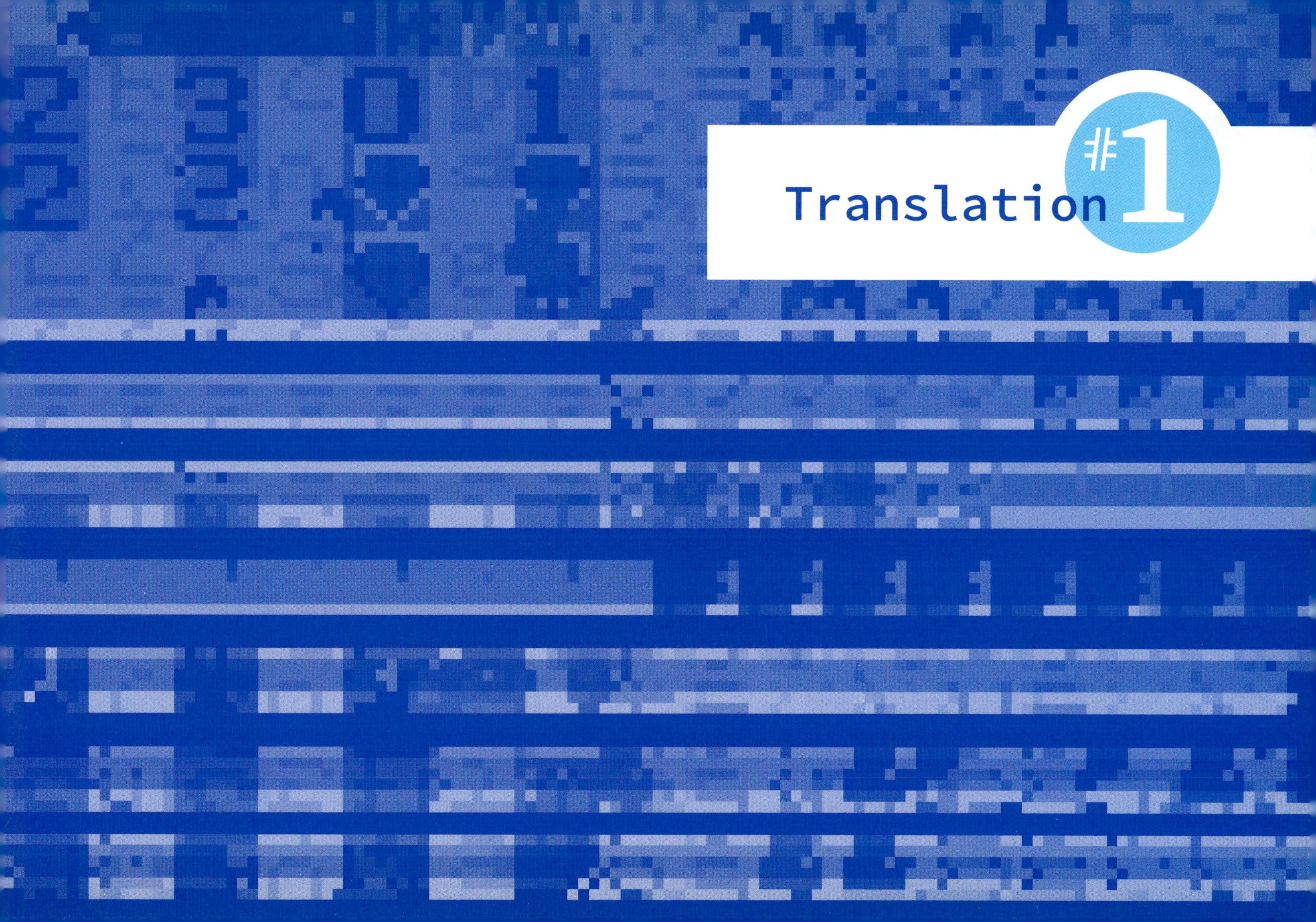
#1
Translation

Translation #1: Traditional Machine Translation

I began work on my game translation tool in late November 2016. A few weeks later, on December 3, I ran *Final Fantasy IV*'s Japanese script through Google Translate and inserted its English translation into the game. This new script, which I now call "Translation #1", serves as a snapshot of "traditional" machine translation. As such, it features many of the qualities associated with machine translations: 100% nonsensical grammar, poor accuracy, and completely untranslated text.

Grammar & Wording

More than anything, Translation #1 can be described as a grammatical mess. Almost every sentence sounds like a cluster of random phrases tied together by gibberish grammar. This form of nonsense is representative of what's technically known as "phrase-based machine translation", in which lines of text are cut into smaller chunks before being translated. If you've ever used an online translation tool, these five examples are probably on par with your own experiences.

The Dark Knight Cecil, now stripped of his position...

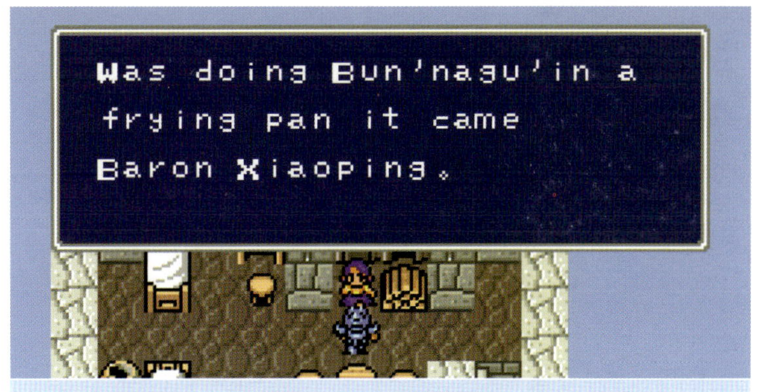

Some Baron soldiers showed up here, but I bashed 'em with my frying pan.

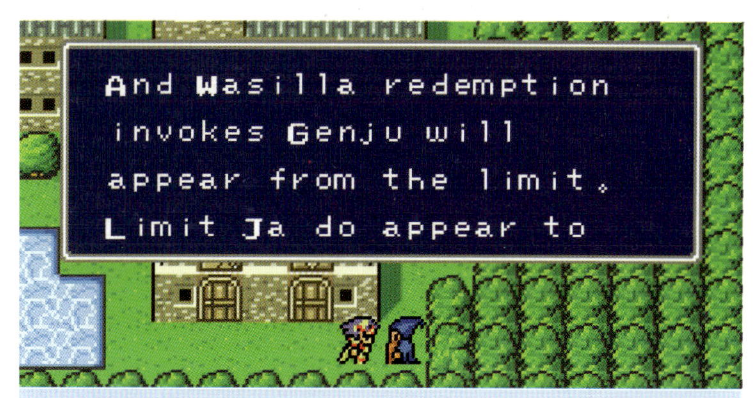

The Eidolons that we summoners call forth appear from the Eidolon Realm. It supposedly exists deep under the earth.

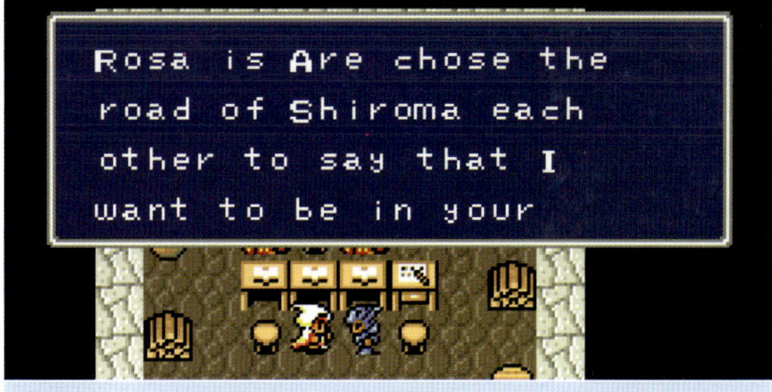

Rosa chose the path of a white mage because she wants to help you.

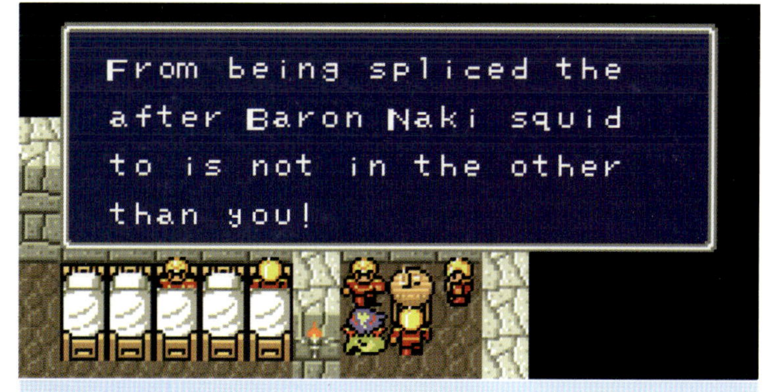

You're the only one who can be the late king's successor!

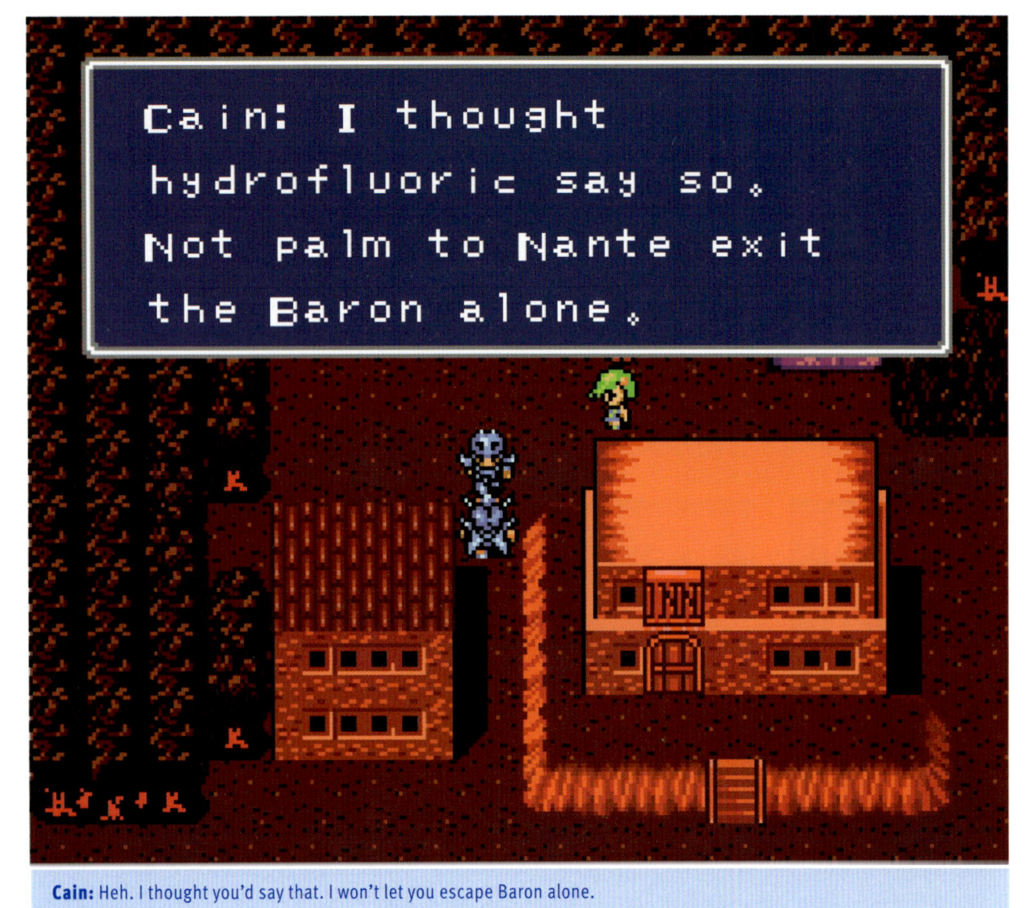

Cain: Heh. I thought you'd say that. I won't let you escape Baron alone.

Purpose & Awareness

Many of the issues in Translation #1 stem from the machine translation system's lack of awareness. Basically, the machine had no idea it was dealing with text from a fantasy-themed video game. It was also unaware that its translation had a specific audience: gamers.

Without this vital information, the machine made all the wrong assumptions and decisions. Instead of looking for fantasy-themed words, it focused on terms from the industrial, medical, and geopolitical fields. We usually think of mistranslations as translation inaccuracies, but Translation #1 demonstrates how a mistranslation can be tone-deaf as well.

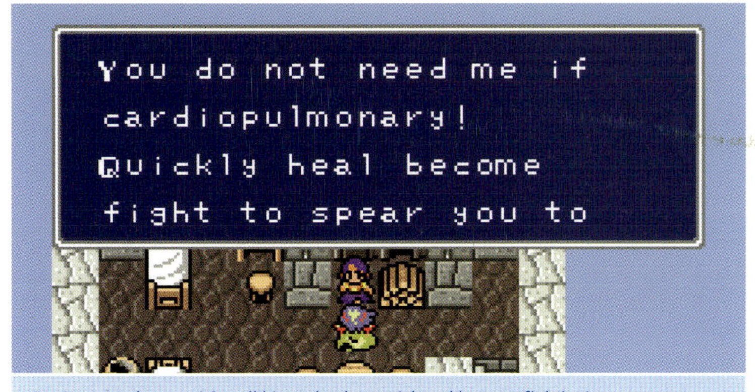

Don't worry about me! Go tell him to heal up quick and keep on fighting!

Palom: We're accompanying Cecil to spy on—

Kluya's Lunarian blood made it even stronger. To think that you two brothers would be made to fight each other...!

But you will never obtain your ultimate strength if you let such emotions control you.

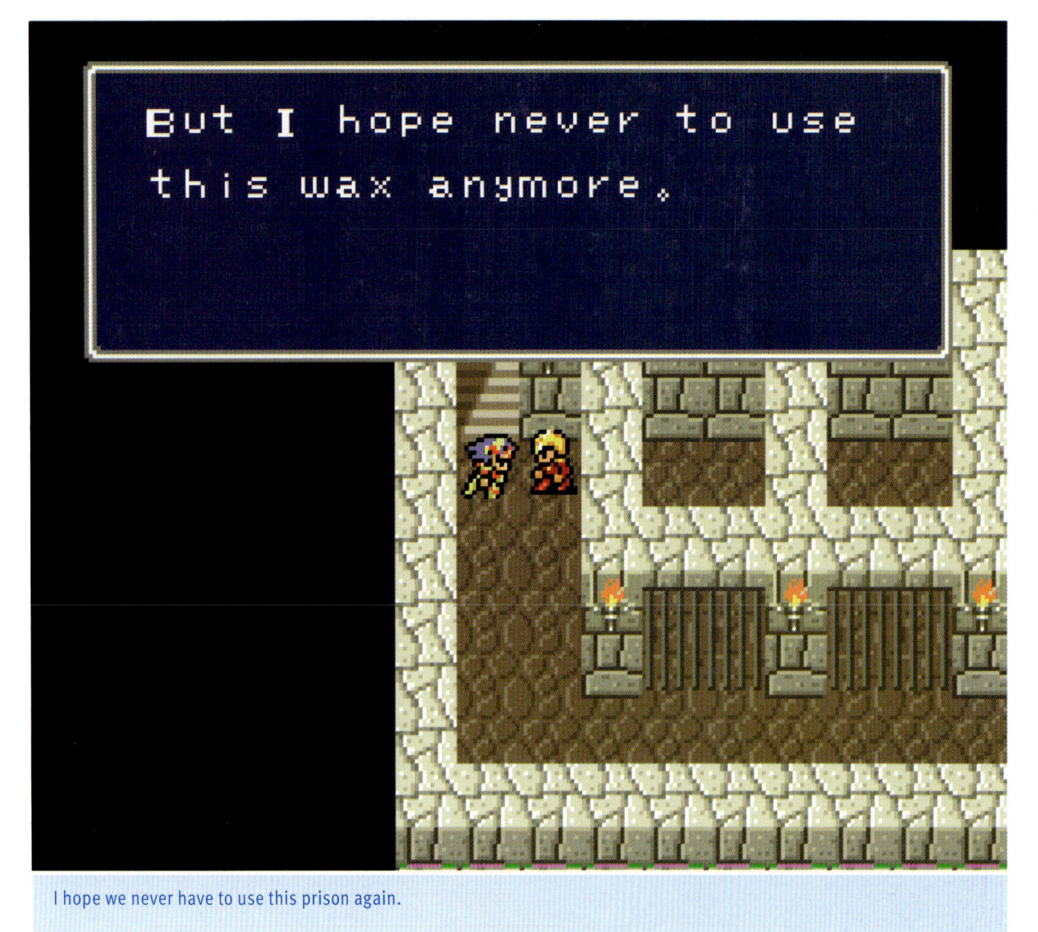

I hope we never have to use this prison again.

Context

The topic of "context" comes up all the time during the translation process. If you're unfamiliar with the term, you can usually just think of "context" as "circumstances". What's going on in this scene? Who's talking? Who's listening? How are they related? Details like these can drastically change the meaning of a line.

Context is also the lifeblood of the Japanese language. For example, once someone has established what they're talking about in Japanese, they'll often leave out subjects, objects, and even verbs from their sentences. So imagine that you walked into the middle of a long Japanese conversation – you'd almost need to be a mind reader to fill in the blanks. That's what it's like to be a translator, and it's why translators always talk about context.

But what happens when the translator is a machine? As Translation #1 shows, the results can be very amusing.

I'm going home now. My daughter's been nagging me lately about how I never leave work!

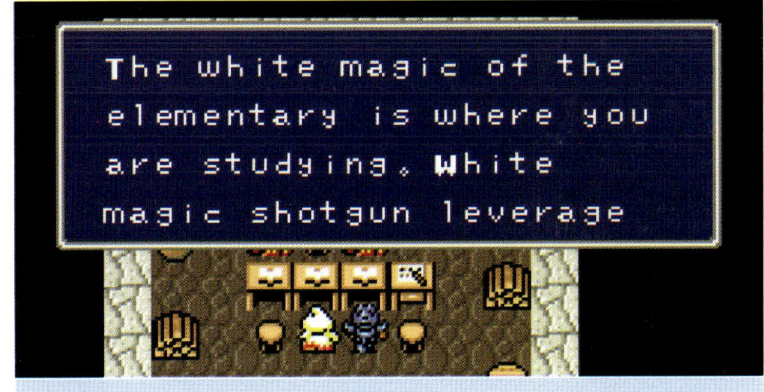

We study basic white magic spells here. Here are the three Cs of white magic...

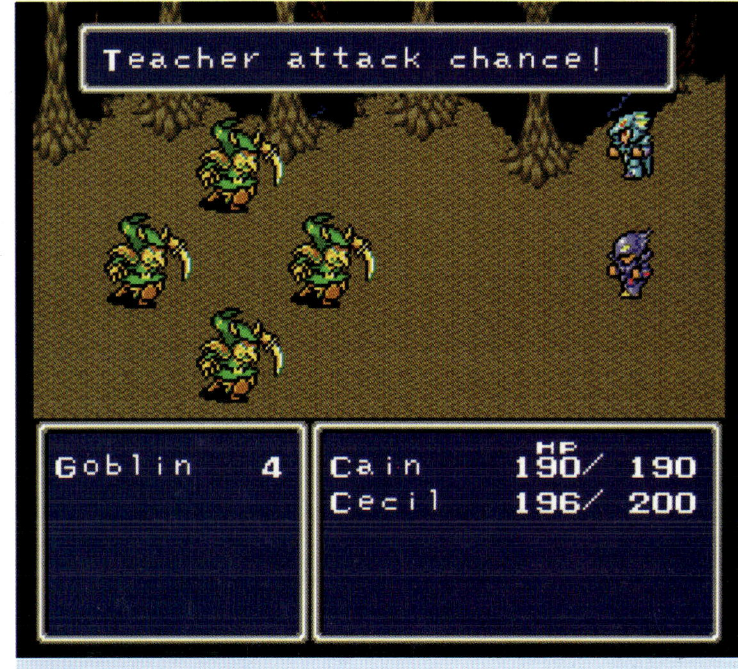

Chance for a preemptive attack!

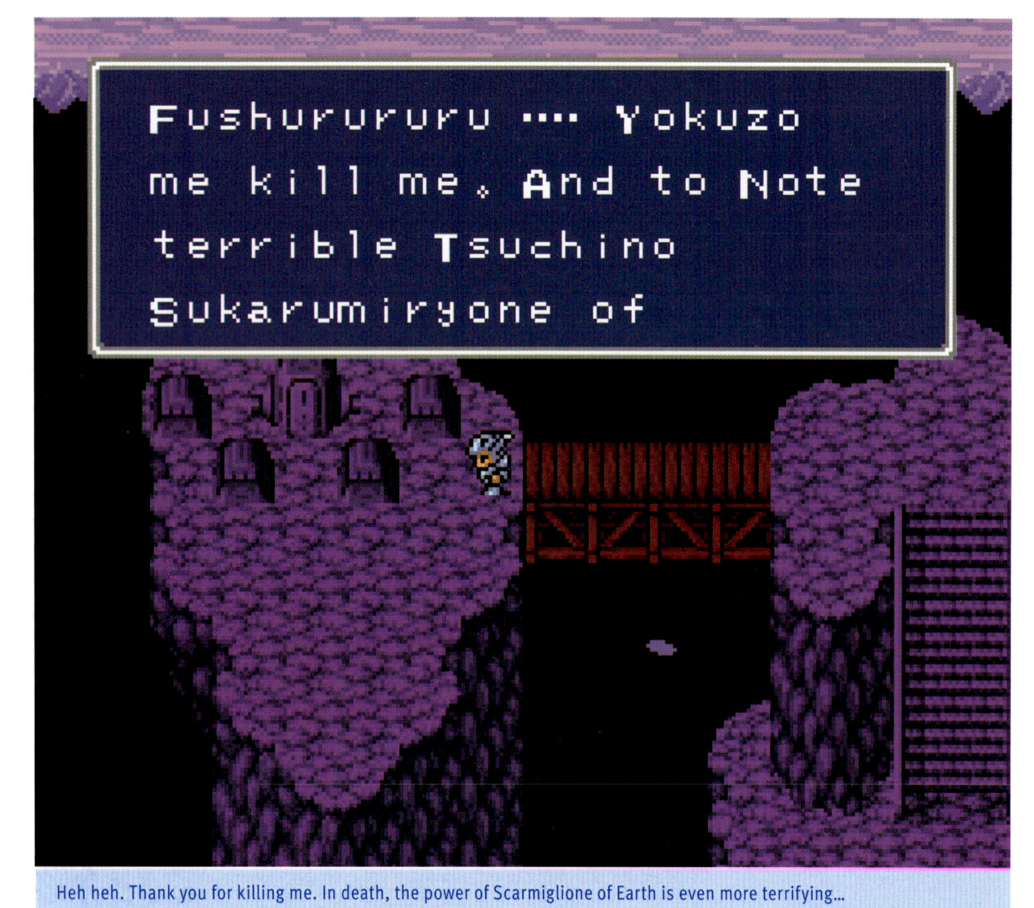

Heh heh. Thank you for killing me. In death, the power of Scarmiglione of Earth is even more terrifying...

Completeness

Translation #1 features a lot of untranslated text, mostly because the machine struggled with the fantasy lingo and each character's unique speaking style. So, instead of translating these problematic phrases, the machine simply wrote out their Japanese pronunciations using the English alphabet. These untranslated bits rarely stretched beyond four words in a row, however – the machine usually got back on track pretty quickly.

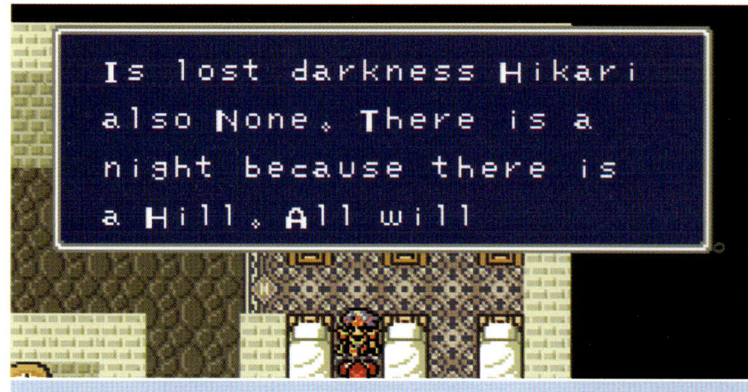

Without darkness there is no light. Because there is day there is also night.

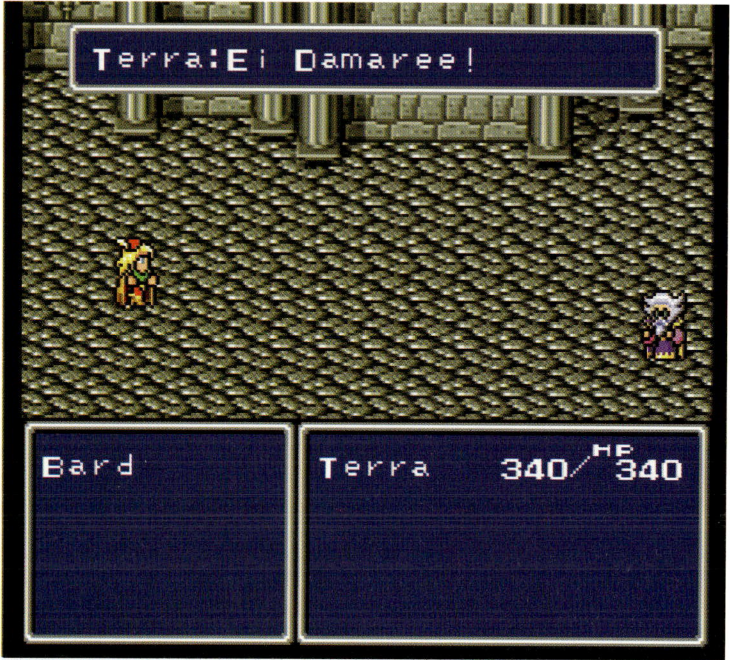

Tellah: Just shut up!

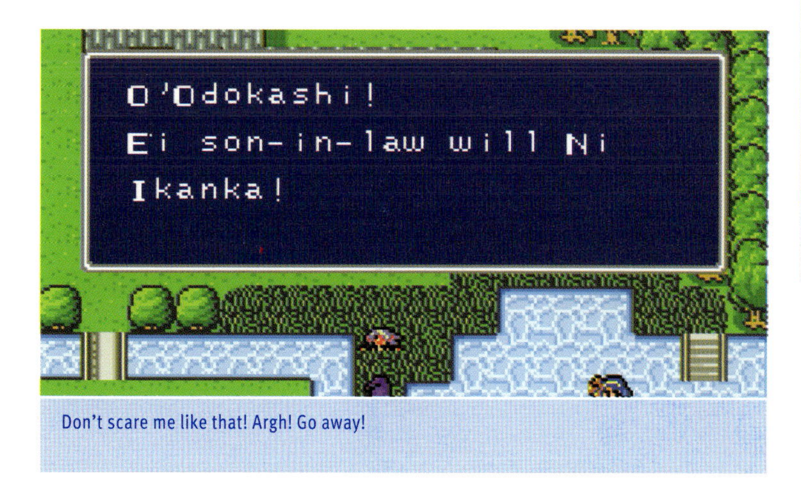

Don't scare me like that! Argh! Go away!

I'm happy... So very happy... that I get to end your lives!

Effectiveness

There's an enormous difference between normal, everyday Japanese and the type of Japanese used in movies, TV shows, games, and the like. This "entertainment Japanese" usually causes machine translators to choke. When proper, grammatically correct sentences *do* appear in entertainment, however, machine translators hit the mark very well.

Final Fantasy IV includes many lines of what could be called "classroom Japanese" – a type of basic, polite Japanese that uses 100% proper grammar. Thanks to this, Translation #1 features a handful of well-translated lines. In fact, despite their clumsy phrasing, many other lines in Translation #1 provide enough information to be understood too.

You used to be that Dark Knight! You look so wonderful now!

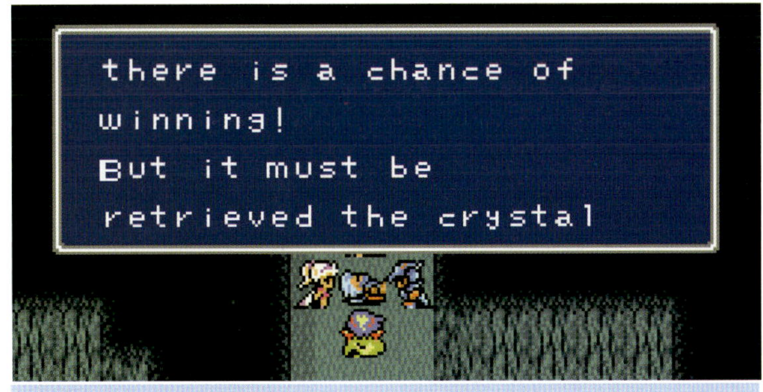

...if we have a chance of winning! But we *must* get the crystals back from them!

The sacred power hidden within you...

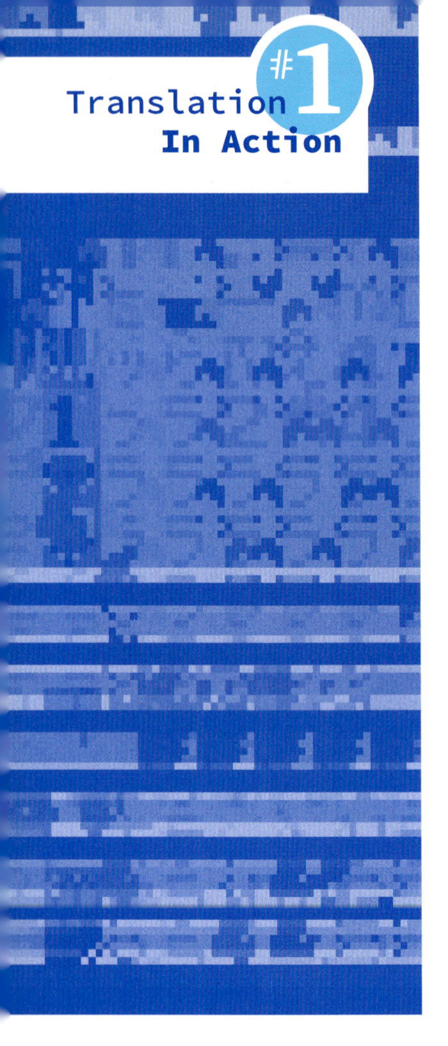
Translation #1
In Action

Soldier: ribbit!
Cecil:? Okay if

Soldier: Gwaah!
Cecil: Are you all right?

Okay?

▲ Tech Support

This dancer ends her text with the Japanese phrase ネッ (*ne*, “Okay?” / “What do you say?”), except it’s presented in *katakana*, which is an unusual choice meant to give the question a little extra flair. Unfortunately, this confused the machine translator into thinking she was about to say ネットワーク (*nettowāku*), the Japanese word for “network”.

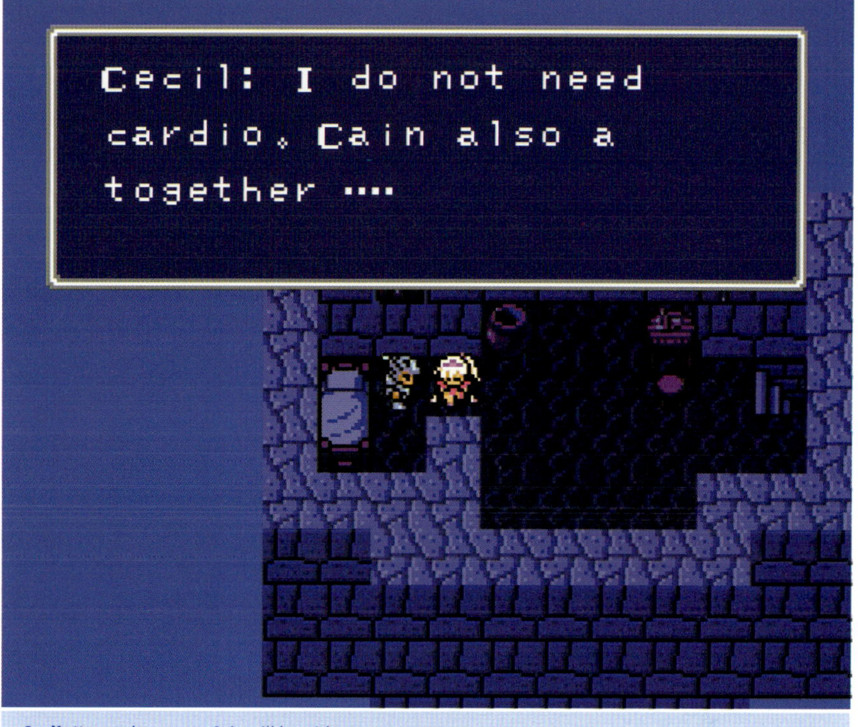

Cecil: No need to worry. Cain will be with me...

Girl: So you're the ones who murdered my mom?!
Cecil: No, we never meant to kill her...
Cecil: I can't believe this is happening...
Cain: I feel bad for this girl, but we'll have to kill her too.

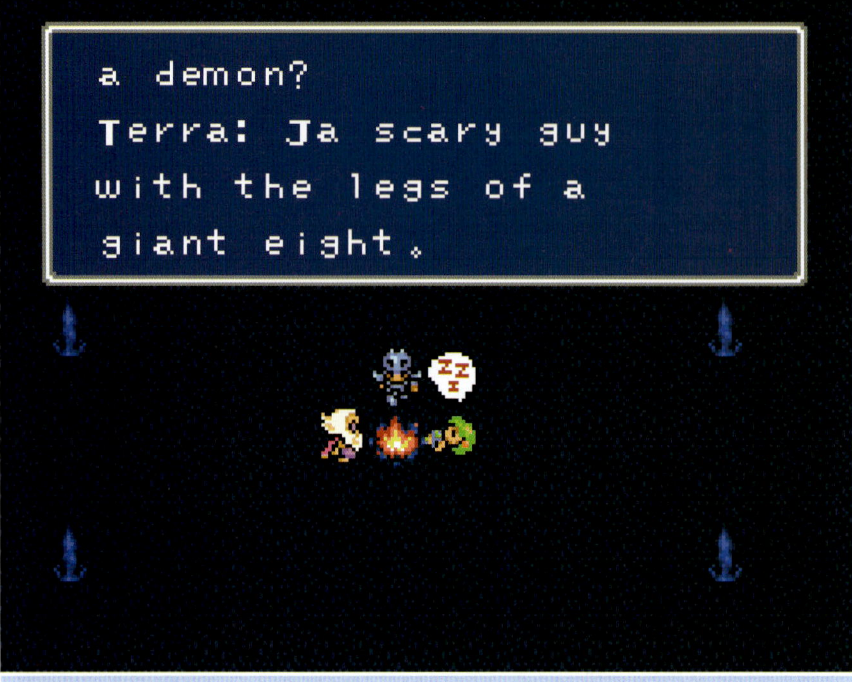

Tellah: It's a fearsome creature with eight giant legs.

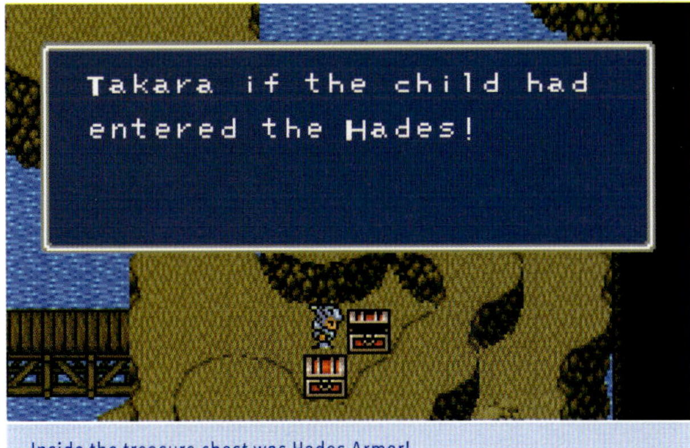

Inside the treasure chest was Hades Armor!

▲ Treasure Trove at Tanagra

Translation #1 is based on an old-school "phrase-parsing" translation system. This approach focuses on individual phrases at a time rather than whole thoughts, so grammar often gets thrown out the window. To make matters worse, a single line of text can be reused countless times throughout a video game. As a result, one little machine-made mistranslation can plague an entire game from start to finish.

Tellah: What do you mean, it's not what I think?!

▲ Sans Silverware

The very first English release of *Final Fantasy IV* is famous for the line "You spoony bard!". The same line is mostly unrecognizable in Translation #1, but for good reason: the original Japanese line was notably different to begin with.

Tellah: I don't need your help! I'll kill Golbez on my own!

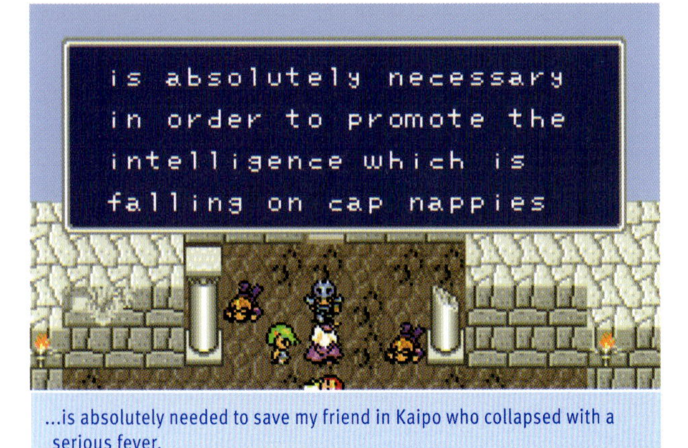

...is absolutely needed to save my friend in Kaipo who collapsed with a serious fever.

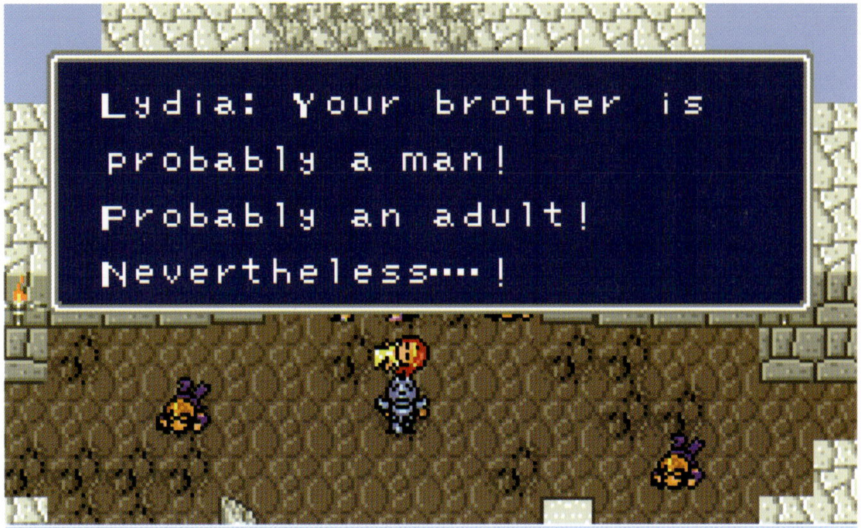

Lydia: Your brother is
probably a man!
Probably an adult!
Nevertheless....!

Rydia: You're supposed to be a man! You're supposed to be a grownup! But, instead, you act like...

▲ What Is a Man?

There's another basic aspect of the Japanese language that can cause translation headaches: pronouns are used far less often in Japanese than in English. For example, instead of using the pronoun "you", it's common to use the listener's name or a familial term like "brother" or "sister". Many times, no word is needed at all. Recognizing unstated, implied information in a sentence – and then using it to make logical decisions – is another obstacle that continues to thwart machine translators.

You were suffering too, I see. I will pray for you as well.

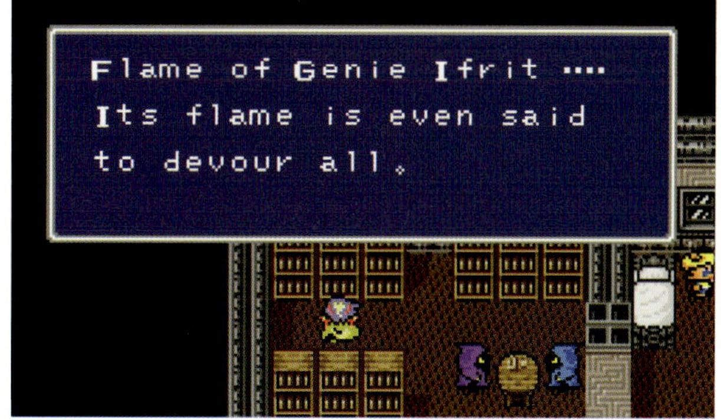

The Djinn of Fire, Ifrit...
His flames are said to burn all into nothing.

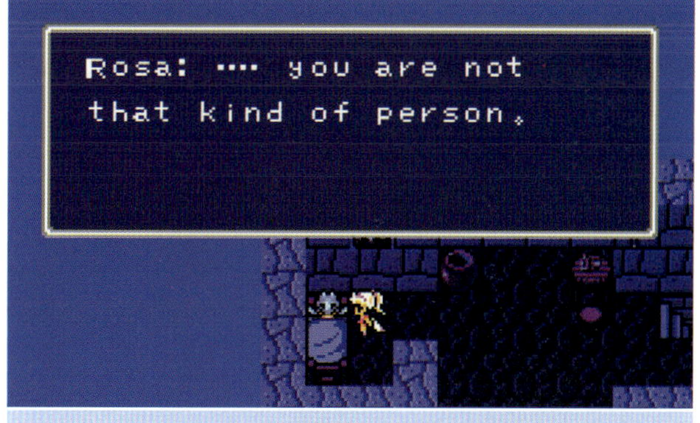

Rosa: ...You aren't that sort of person.

There is Hope

As we've already seen, not everything in Translation #1 is nonsense. The three screenshots on this page offer great examples of how machines thrive when given proper, standard Japanese to translate.

Even if a machine translation looks okay at first glance, though, it can be risky to use it as-is. Some of the sneakiest problems hide in plain sight, as seen in the image above.

◂ On the Menu

Although the machine translation process left most of the game's menu text intact, it also produced a handful of amusing English names. Unfortunately, most of these names were so long that they had to be heavily trimmed to fit inside cramped menu boxes, so here are some of the full-length highlights:

- The **Engineer** class name – a title only held by a crusty old mechanic named Cid – became **Sister-in-law**
- The **Protes** white magic spell became **Protestant**
- The **Smoke Bomb** ninja skill became **Smoke Lumps**
- The powerful **Crystal Shield** became **vertical of Crystal**
- The **Triangular Hat** or **Tricorne Hat** became **participation hat**
- The **Bacchus' Wine** item became **Salmon of Bacchus**
- The **Lunar Curtain** item became **Curtain of the month**
- The **Thunder Dragon** became **Leprosy flow**
- The **Black Magic Laboratory** became **Chroma consecration moisture absorption**
- The frightening **Death Sentence** status ailment became **Sentence of teeth**
- The *Final Fantasy* series' famous **Zantetsuken** instant death attack became **Remaining in soap**
- The **Float-Eyeball** enemy became **Flow Thailand ball**

We're digging a tunnel into the Tower of Babil to avenge our king and queen.

Lugae took it upon himself to transform the king and queen into monsters.

The king and queen regained control of their minds!

▲ Queasy Queen

In Translation #1, the queen of the ninja kingdom became the "Vomiting Queen". This mistake happened because the machine translator mistook the phrase *ō to* ("the king and") for a word that's pronounced the same: *ōto* ("vomit"). As a result of this basic mistake, the "Vomiting Queen" mistranslation remains consistent throughout the game.

Additionally, the machine tripped over *ō to* so much that it accidentally gave the queen a name: Nikki. She doesn't have a given name in the original Japanese script, though, and there's nothing even close to "Nikki" in the Japanese line. So where did the word come from? It's a machine translation mystery.

▸ Potty Mouth Machine

Translation #1 features many lines of text with surprisingly crude wording. There's something charming about seeing a machine try to create swears and insults, and this machine translation didn't disappoint.

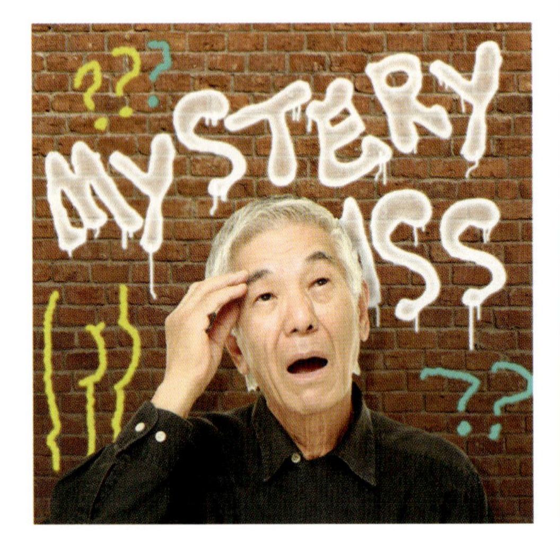

Old Man: If you're aware of that, then you should stop chasing after girls all the time!

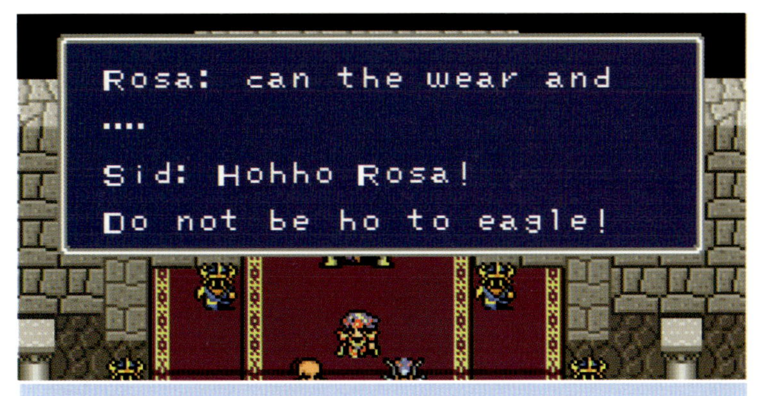

Rosa: Be careful...
Cid: Ho-ho! Rosa! Don't fall for me now, ya hear!

Palom: They're...!

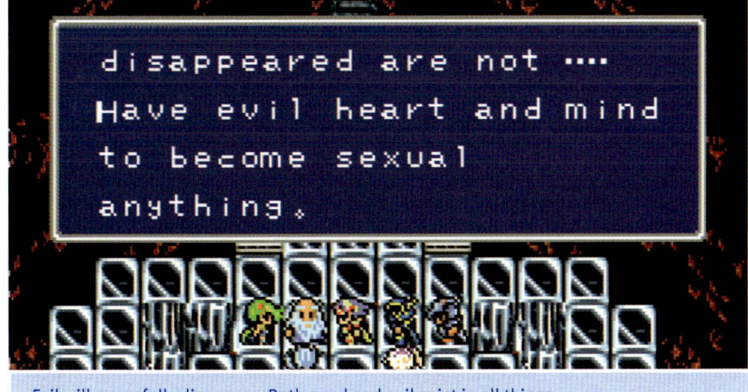

Evil will never fully disappear. Both good and evil exist in all things.

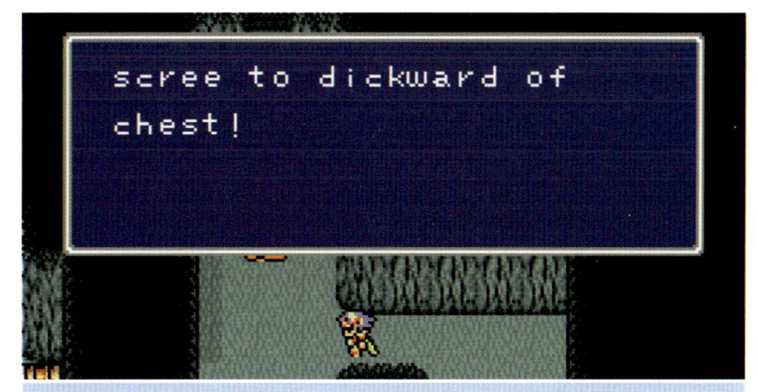

Maybe this will help you remember!

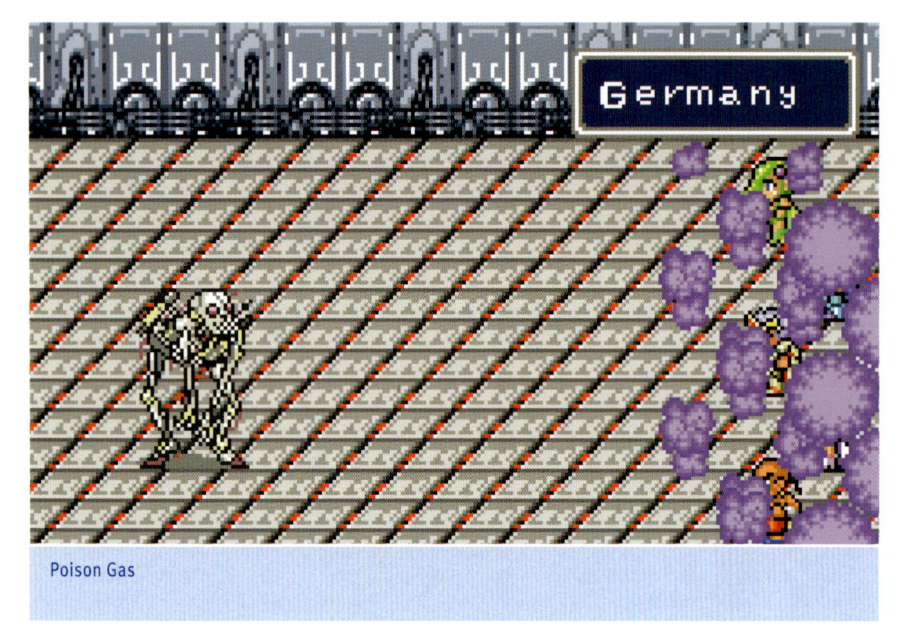

Poison Gas

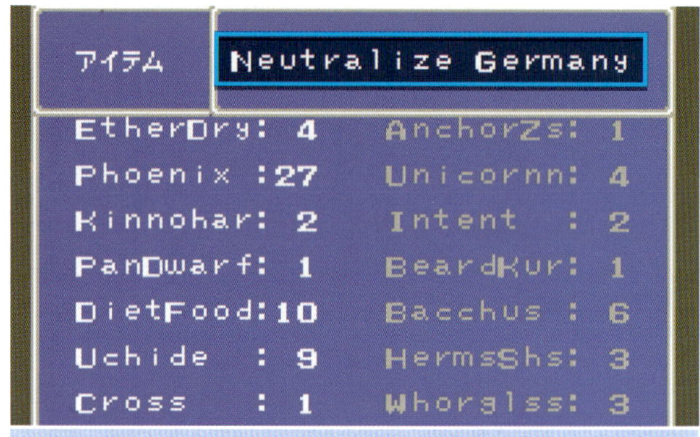

Neutralize poison

Poisoned!

▲ German Infatuation

The Japanese word *doku* has many meanings. For example, it can potentially mean "poison", "reading", "solitude", "to move out of the way", or "skull". It's also often used as a Japanese abbreviation for the name "Germany". It's clear from context that it's supposed to mean "poison" throughout *Final Fantasy IV*, but the machine behind Translation #1 was unaware that it was translating a video game. The result was a very wrong translation choice!

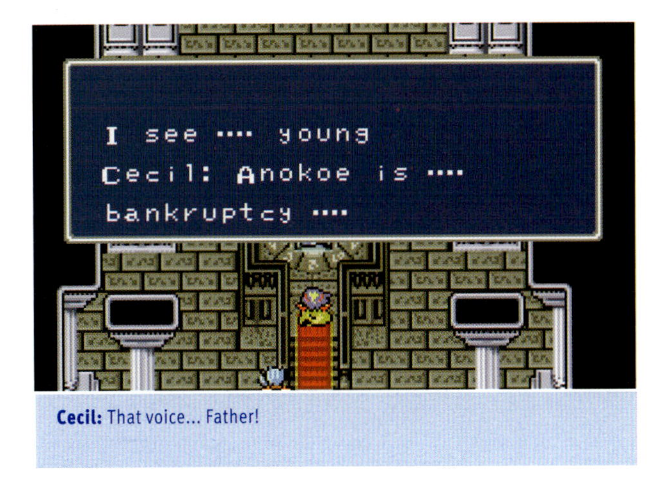

Cecil: That voice… Father!

▲ Suddenly Broke

In everyday conversational situations, the Japanese word *tōsan* can mean "father". In business contexts, it instead means that a company has gone bankrupt. Of course, corporate bankruptcy isn't a theme commonly explored in Japanese fantasy RPGs, but machine translators don't know that.

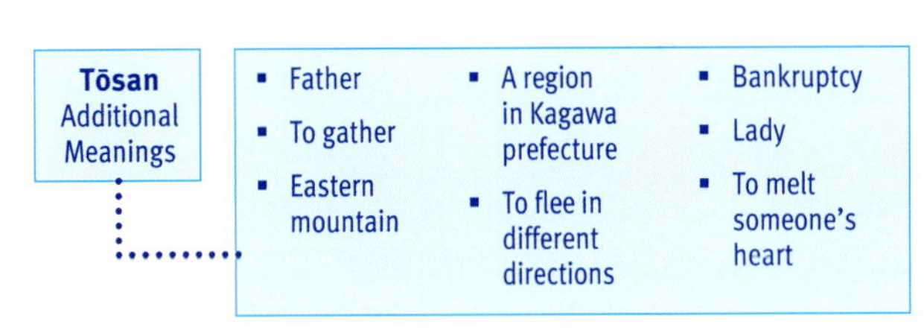

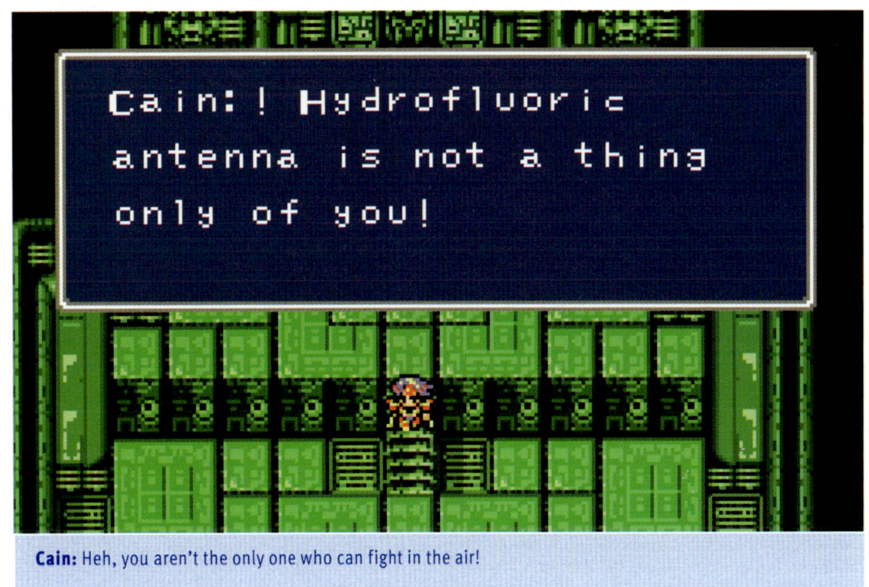

Cain: Heh, you aren't the only one who can fight in the air!

▲ Poor Reception

In Translation #1, Cain's cocky laughter is consistently replaced with technical words from the chemical industry. In this instance, his laugh is translated as "hydrofluoric". What's more, the word *sentō* ("battle") appears to have been mistranslated as "antenna" and the word *kūchū* ("in the air", "aerial") seems to have disappeared entirely.

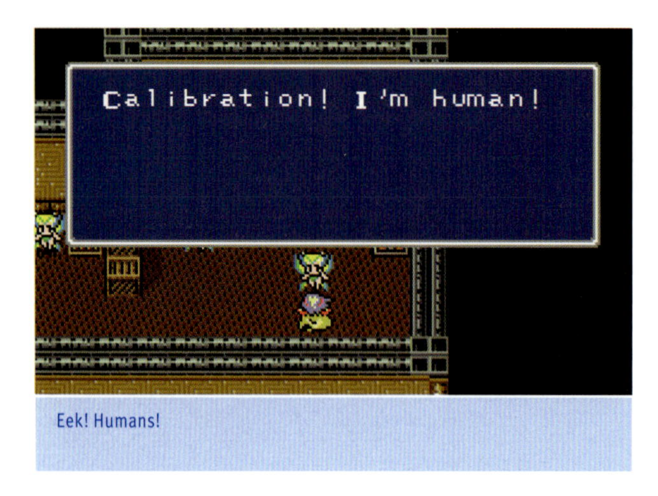

Eek! Humans!

▲ Magical Cyborg

This fairy's second sentence lacks a subject in Japanese, so it's understandable that the machine translator would have trouble filling in the blanks.

So how does "calibration" fit into the picture? As far as I can tell, it seems that the machine mistook the word *kya* (a shrieking sound) as an abbreviation of *kyaribureshon* ("calibration"). Whatever the reason, this mistake is so ridiculous that it's actually kind of mesmerizing.

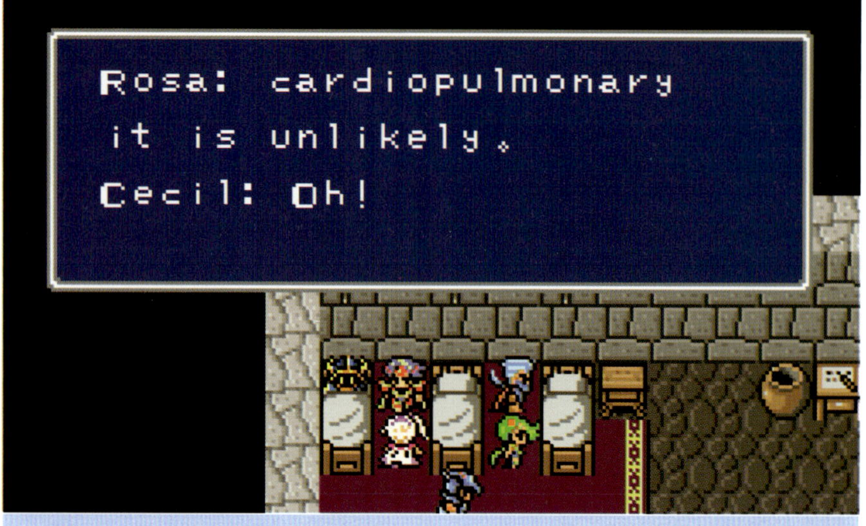

Rosa: It looks like we have nothing to worry about.
Cecil: Indeed!

▲ White Mage Diagnosis

The Japanese word *shinpai* ("worry") is an extremely common term that's used all the time in real life and in entertainment. In the medical field, however, *shinpai* can also mean "the heart and lungs", which is why Translation #1 is filled with the word "cardiopulmonary". This is a great example of a machine translator being unaware of its purpose *and* ignoring context. After all, it's a million times more likely that someone will use the phrase "there's no need to worry" rather than "there's no need for the heart and lungs".

We'll avenge her!

Cecil: It's right here! Now where is Rosa?!
Cain: Heh. Just calm down.

Cain: We have to destroy it!

▲ Rough Neighborhood

The Japanese version of this line uses the word *sento*, which is a non-standard, entertainment-style way of saying "have to". As a noun, *sento* can also mean "saint", which is often abbreviated as "St." in English. Not only did the machine make the wrong translation choice here, the position of "St." makes it look like it's an abbreviation for "Street" as well. This is a perfect example of how a single mistranslation can lead to misunderstandings that stray farther and farther from the original text.

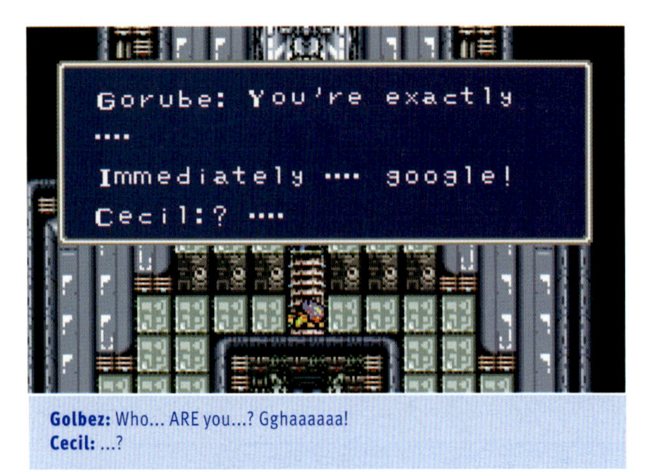

Golbez: Who... ARE you...? Gghaaaaaa!
Cecil: ...?

Still, how pathetic he is! Go wallop him with this to put some life back into him!

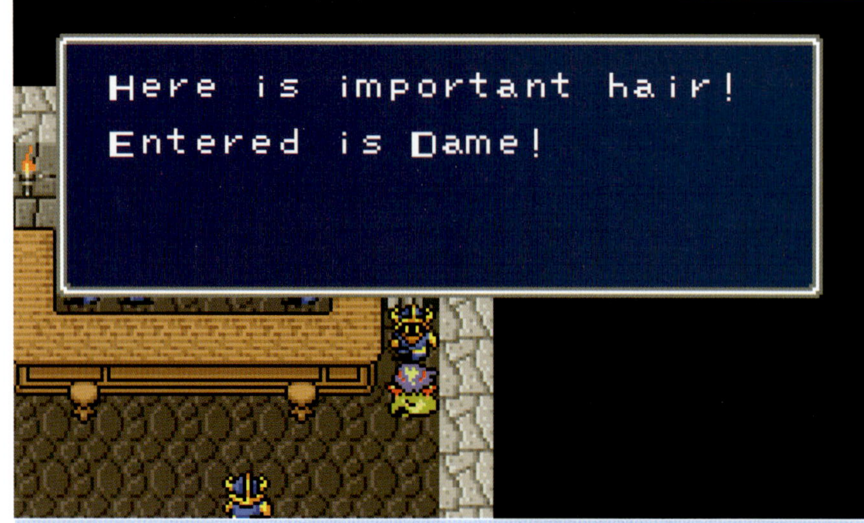

This is an important room! You can't go in!

▲ Cutting Remarks

The dwarves in *Final Fantasy IV* speak in a unique way in Japanese: they tend to hold their vowels at the end of sentences. This speech quirk confounded the machine translator, which mistook the word *heya* ("room") in this line for the word *heyā* ("hair").

▶ First Blood Part IV

The Japanese word *ranbō* means something like "violent" or "rowdy". By sheer coincidence, it's also the Japanese spelling of the name "Rambo", which explains this surprising reference in Translation #1. Incidentally, at one point in *Final Fantasy IV*, Cid is tough enough to strap a bomb to himself and blow it up as he plummets to his death – yet he survives it all just fine. So even though it was a mistranslation to call him "Rambo" here, it's certainly a suitable nickname!

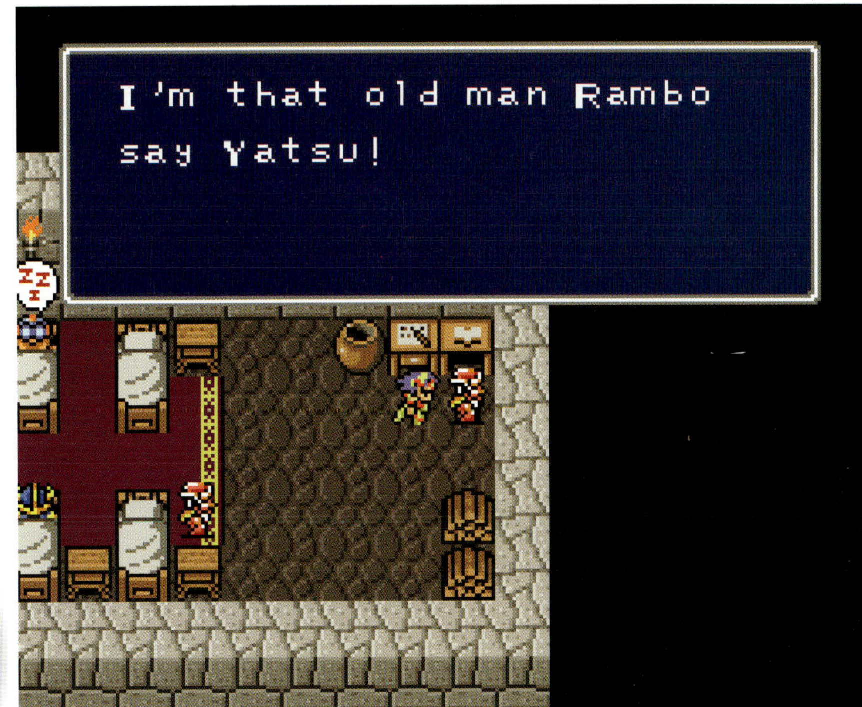

That older fellow is rowdy, but he's a good guy!

Rambo is definitely a *ranbō* kind of guy

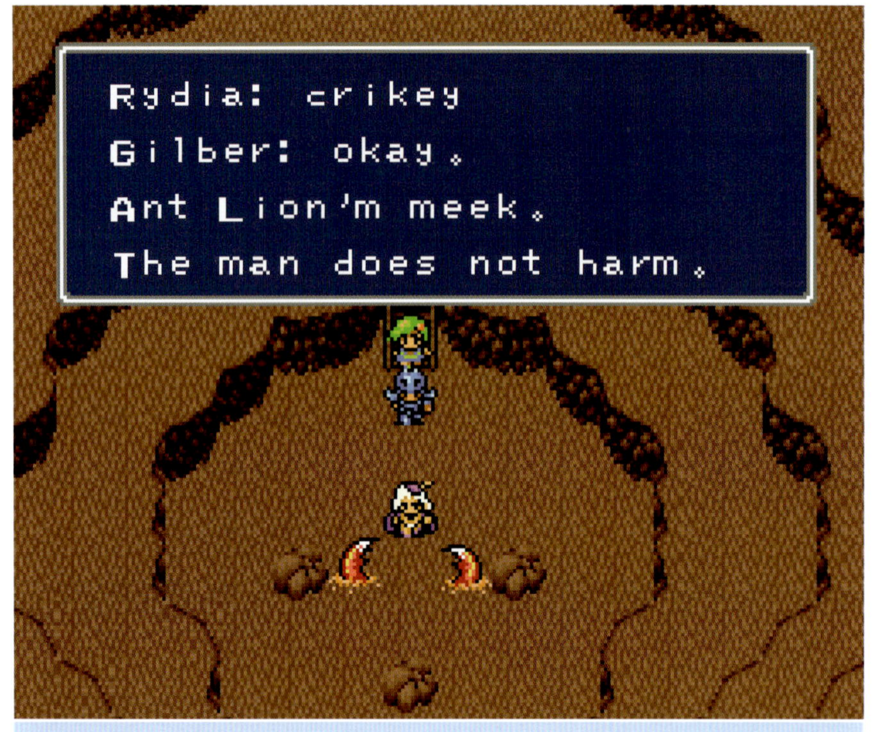

Rydia: Eek!
Gilbert: It's okay. The Antlion is tame. It doesn't harm people.

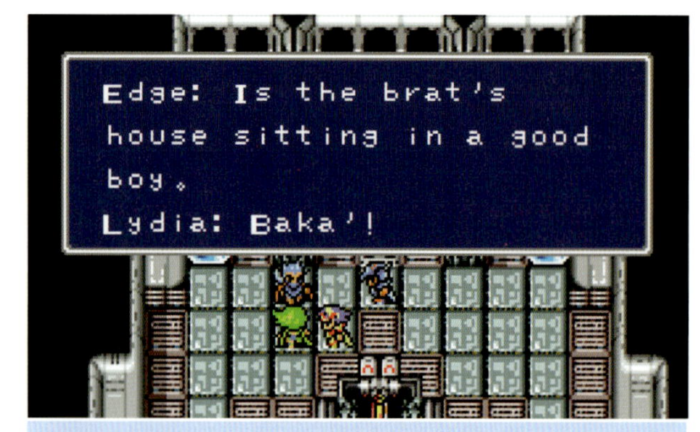

Edge: Okay, now be a good girl and stay put while we're gone.
Rydia: You idiot!

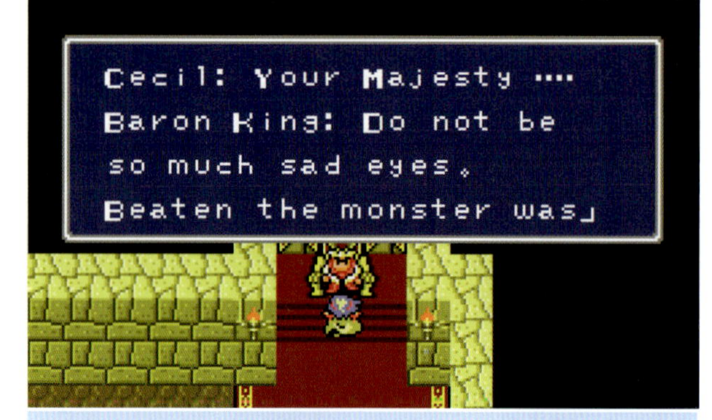

Cecil: Your Majesty... **King of Baron:** Do not look so sad. Although I was slain by a monster, I also gained everlasting strength.

This is the rescue team's recovery room.

◂ Geography Bee

The Japanese noun suffix *shitsu* is often used to mean "room" – for example, a "boiler *shitsu*" is a "boiler room" and a "game *shitsu*" is a "game room". However, the word *shitsu* has another, completely separate meaning: "quality". In this line from Translation #1, we see that the machine translator made the wrong choice, resulting in the phrase "nursing quality" rather than "nursing room".

A similar misunderstanding turned *kyūjo tai* ("rescue team") into a phrase about Thailand. This happened because *tai* is the Japanese name for Thailand. Unfortunately, there are many unrelated Japanese words that coincidentally have *tai* in them. As a result, Translation #1 includes a surprising number of references to Thailand.

In simpler terms, imagine you were translating from English to Spanish and decided to replace every instance of "hat" to "sombrero" – including changing "that" to "tsombrero" and "chatty" to "csombreroty". Only a bonehead translator would make a mistake that basic!

Cecil: It's in here?
Cid: It's like they say, "To really hide something, put it in the lighthouse!"

◂ Academic Waiver

In Japanese, Cid uses a fancy phrase to explain that he hid his airship where no one would ever find it: right under everyone's noses. In Translation #1, he babbles something about the University of Tokyo. This mistranslation happened because the word *tōdai* can mean "lighthouse", but it also happens to be the abbreviation for *tōkyō daigaku* ("University of Tokyo"). It's an understandable mistake, but I'm fairly sure Cid never knew about real-life Tokyo.

Cecil: I-it's collapsing! **Cain:** Damn! **Cid:** Waaah!

Go, my lovelies...

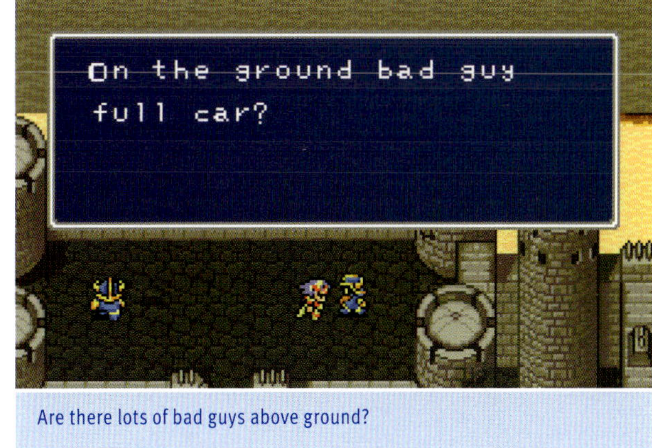

Are there lots of bad guys above ground?

▲ It's the HOA's Fault

Although machine translators struggle with major issues like context and purpose, they also have trouble with simpler things like speech stutters. In this example, the phrase *ku kuzureru* ("I-it's collapsing!") was mistaken for two separate words: *ku* ("ward", "district") and *kuzureru* ("to collapse"). The meaning of the original line can still be understood in Translation #1, but this one mistake keeps the scene's translation from being flawless.

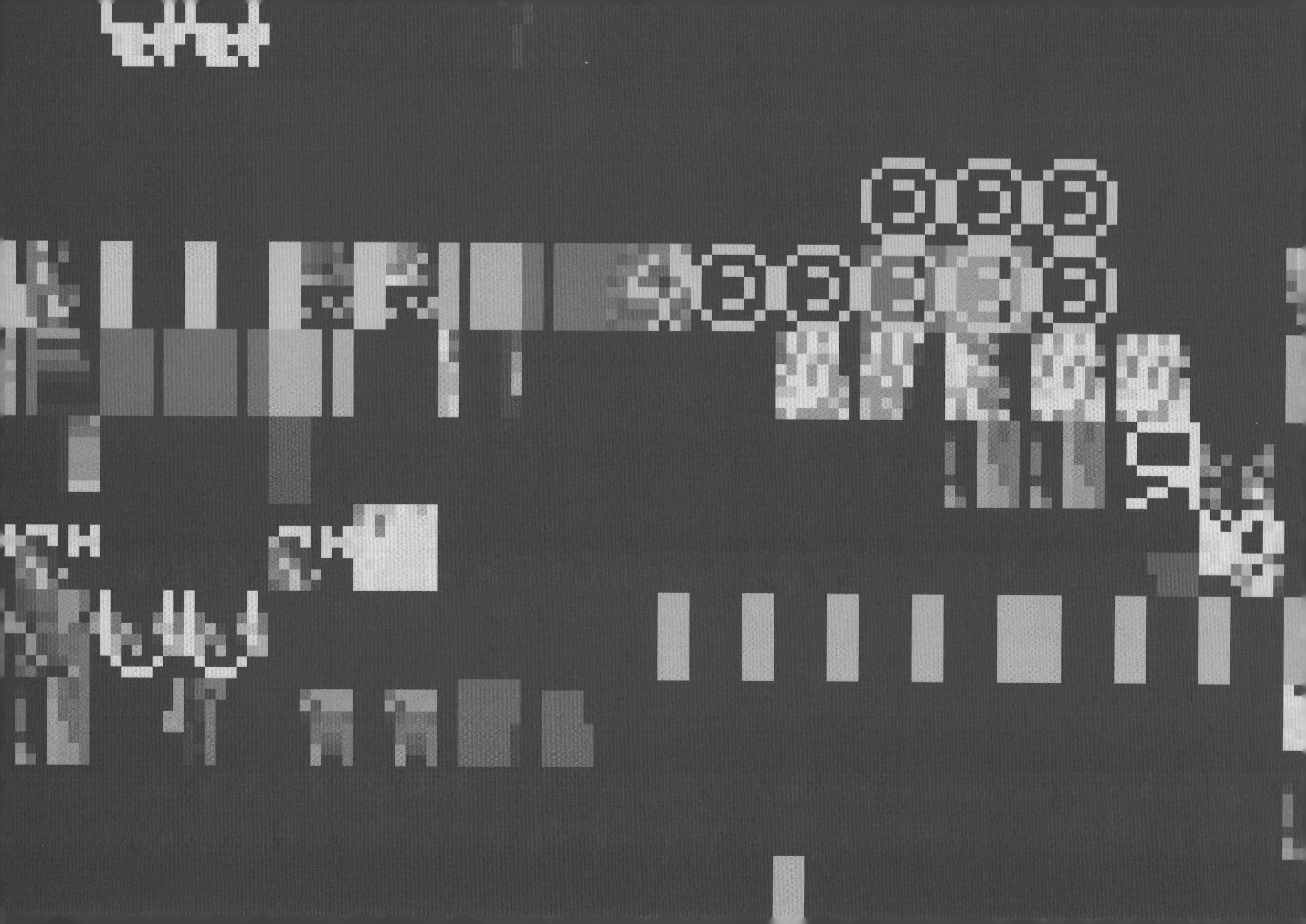

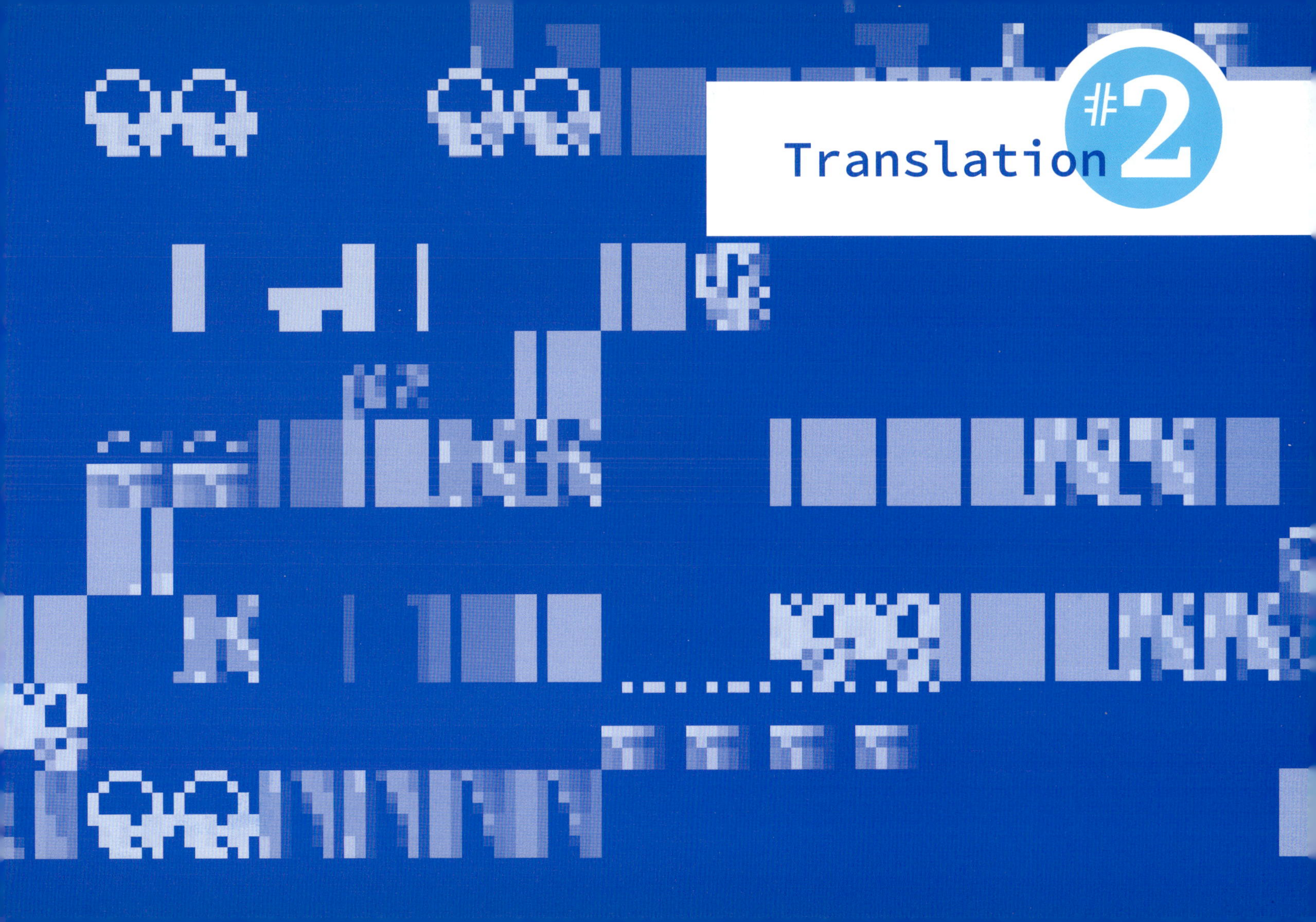
Translation
#2

Translation #2: Cutting-Edge Neural Network Translation

On December 6, 2016, I ran *Final Fantasy IV*'s Japanese text through Google Translate for a second time. To my surprise, the translation it spit out was completely different from the one it had given me three days earlier! As I later learned, Google had upgraded its system to use a newfangled technology known as "neural network machine translation".

This new *Final Fantasy IV* translation – which I creatively called "Translation #2" – was nothing like the previous machine translation. It remained wildly inaccurate, but this time the English grammar was actually good! If Translation #1 sounded like it was written by a cat stepping on a keyboard, then Translation #2 sounded like it was written by a drunkard having a nightmare.

In this chapter, we'll explore Translation #2 from start to finish. We'll also use Google's neural network machine translator – which is still online as I write this – to investigate why certain phrases in Translation #2 wound up the way they did. These details will help us peer into the mind of the newest machine translation technology.

I was so surprised and excited when I realized this second translation was completely different. Translation #2 turned out to be a treasure trove of translation issues I'd never seen before!

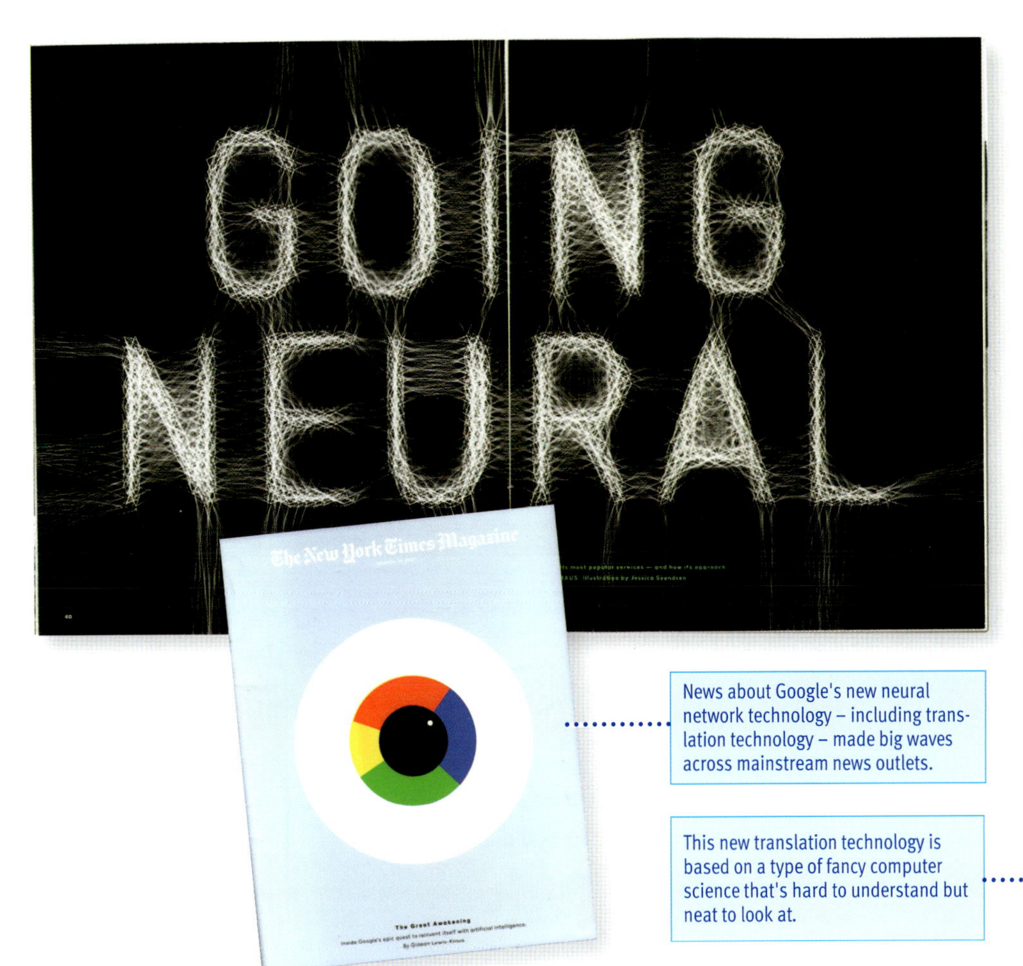

News about Google's new neural network technology – including translation technology – made big waves across mainstream news outlets.

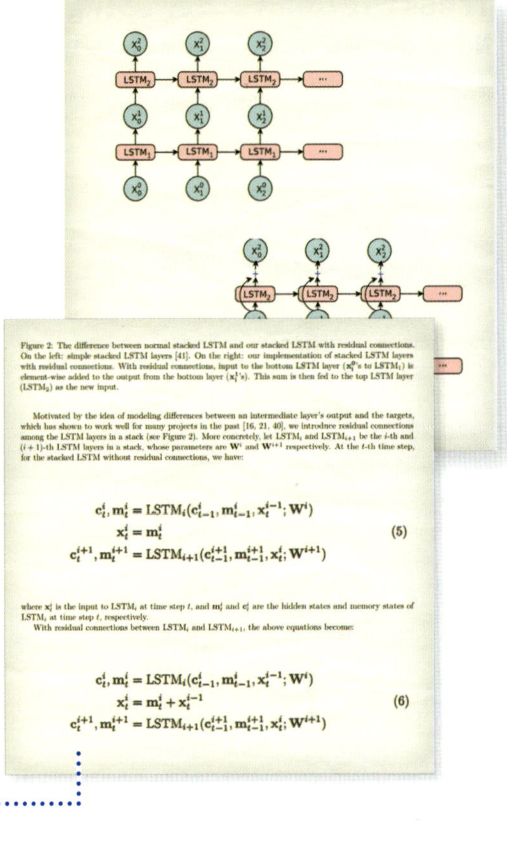

Figure 2: The difference between normal stacked LSTM and our stacked LSTM with residual connections. On the left: simple stacked LSTM layers [41]. On the right: our implementation of stacked LSTM layers with residual connections. With residual connections, input to the bottom LSTM layer ($\mathbf{x}_i^0$'s to LSTM_1) is element-wise added to the output from the bottom layer ($\mathbf{x}_i^1$'s). This sum is then fed to the top LSTM layer (LSTM_2) as the new input.

Motivated by the idea of modeling differences between an intermediate layer's output and the targets, which has shown to work well for many projects in the past [16, 21, 40], we introduce residual connections among the LSTM layers in a stack (see Figure 2). More concretely, let LSTM_i and LSTM_{i+1} be the i-th and $(i+1)$-th LSTM layers in a stack, whose parameters are $\mathbf{W}^i$ and $\mathbf{W}^{i+1}$ respectively. At the t-th time step, for the stacked LSTM without residual connections, we have:

$$\begin{aligned} \mathbf{c}_t^i, \mathbf{m}_t^i &= \text{LSTM}_i(\mathbf{c}_{t-1}^i, \mathbf{m}_{t-1}^i, \mathbf{x}_t^{i-1}; \mathbf{W}^i) \\ \mathbf{x}_t^i &= \mathbf{m}_t^i \\ \mathbf{c}_t^{i+1}, \mathbf{m}_t^{i+1} &= \text{LSTM}_{i+1}(\mathbf{c}_{t-1}^{i+1}, \mathbf{m}_{t-1}^{i+1}, \mathbf{x}_t^i; \mathbf{W}^{i+1}) \end{aligned} \tag{5}$$

where $\mathbf{x}_t^i$ is the input to LSTM_i at time step t, and $\mathbf{m}_t^i$ and $\mathbf{c}_t^i$ are the hidden states and memory states of LSTM_i at time step t, respectively.

With residual connections between LSTM_i and LSTM_{i+1}, the above equations become:

$$\begin{aligned} \mathbf{c}_t^i, \mathbf{m}_t^i &= \text{LSTM}_i(\mathbf{c}_{t-1}^i, \mathbf{m}_{t-1}^i, \mathbf{x}_t^{i-1}; \mathbf{W}^i) \\ \mathbf{x}_t^i &= \mathbf{m}_t^i + \mathbf{x}_t^{i-1} \\ \mathbf{c}_t^{i+1}, \mathbf{m}_t^{i+1} &= \text{LSTM}_{i+1}(\mathbf{c}_{t-1}^{i+1}, \mathbf{m}_{t-1}^{i+1}, \mathbf{x}_t^i; \mathbf{W}^{i+1}) \end{aligned} \tag{6}$$

This new translation technology is based on a type of fancy computer science that's hard to understand but neat to look at.

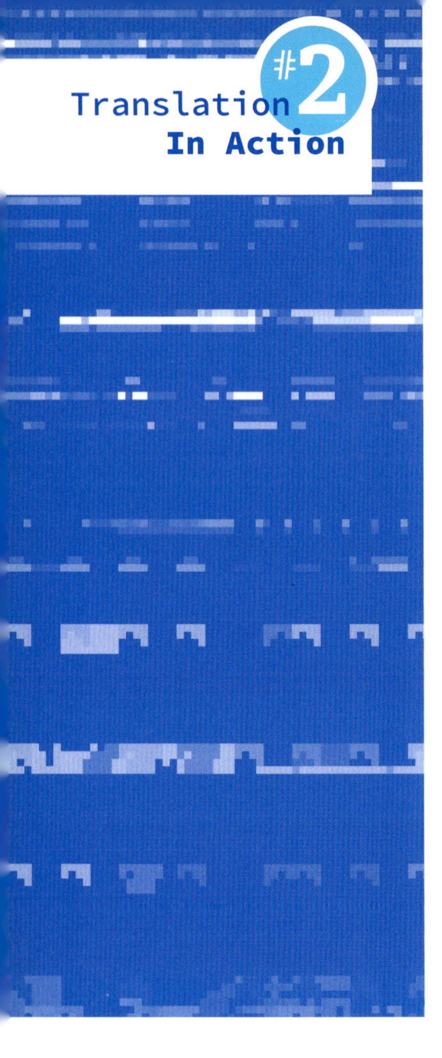

Float-Eyeball

ORIGINAL NAME
Float-Eyeball

Machine Translation

UNSHORTENED NAME
Flow Tie Ball

Text Compression

SHORTENED NAME
FlowTiBl

▲ Knotty Translation

A group of monsters called "Float-Eyeballs" attack the main character during *Final Fantasy IV*'s prologue. Unfortunately, the machine translator misinterpreted "Float-Eyeball" and broke it apart in the wrong place, resulting in an entirely new name: Flow Tie Ball. To make matters worse, because "Flow Tie Ball" was over eight letters in length, the name was automatically compressed into something even less understandable.

Cecil: Is everyone okay?

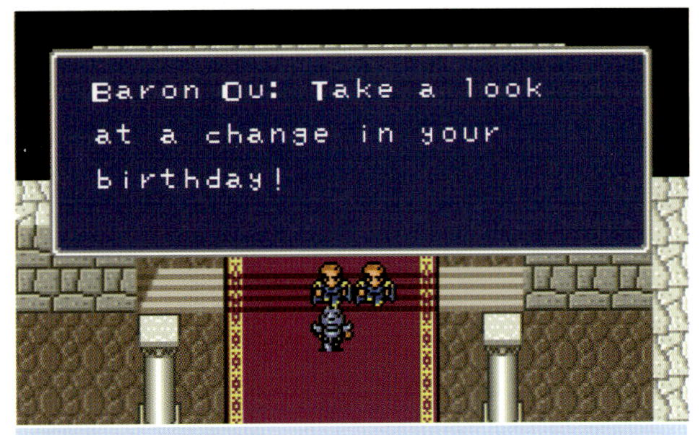

King of Baron: Instead, you are hereby tasked with slaying an Eidolon!

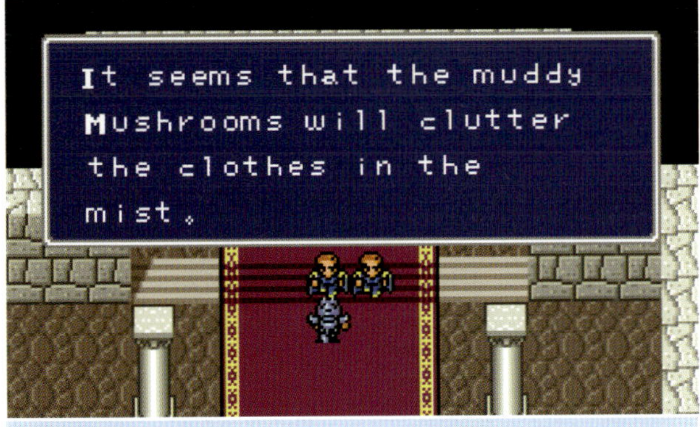

They say a mystical monster Eidolon has appeared near Mist.

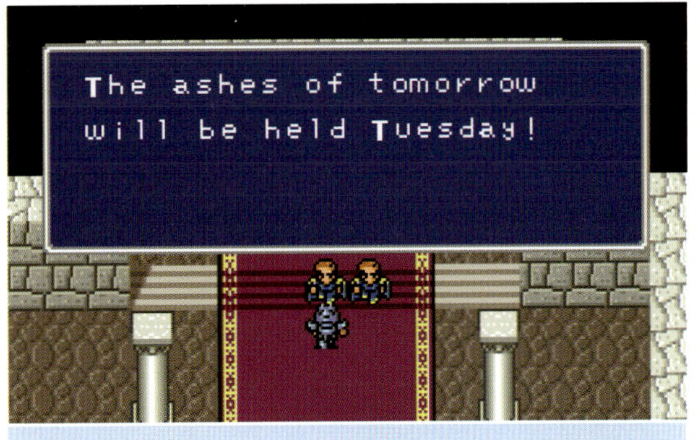

You will depart tomorrow morning!

Story 

Cecil, the captain of Baron's airship fleet, plunders a town and steals its magical crystal. He feels guilty about doing this, however.

Filled with regret, Cecil reluctantly gives the crystal to the king of Baron. The king punishes him for his disloyalty and sends him to Mist Village.

Mato Says 

"Eidolons" are special monsters that can be summoned during battle to attack enemies.

Later in the game, it's learned that the Eidolons live together in a secluded realm underground.

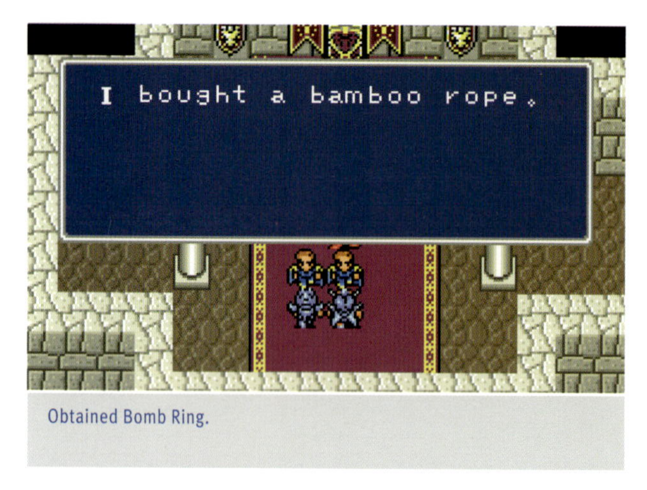

Obtained Bomb Ring.

His Majesty is furious.

▲ Look to the Beginning

This mistranslation features a pattern unique to Translation #2: the machine often takes the first letter of a translated word and inserts a different word that starts with the same letter. In this case, the correct translation of "Bomb Ring" transformed into "bamboo rope". I thought this was pure coincidence at first, but it marked the start of what I call the "First Letter Pattern".

▲ Royal Boundaries

The word *heika* ("His Majesty", "Your Majesty") is consistently mistranslated throughout Translation #2. In this instance, the machine translator ignored the *ka* and translated the word *hei* ("fence") instead. Incidentally, Translation #1 handled *heika* just fine most of the time, so this marks a step backwards in quality. "Gorp" is a little-known English word that can mean "trail mix" or "to devour".

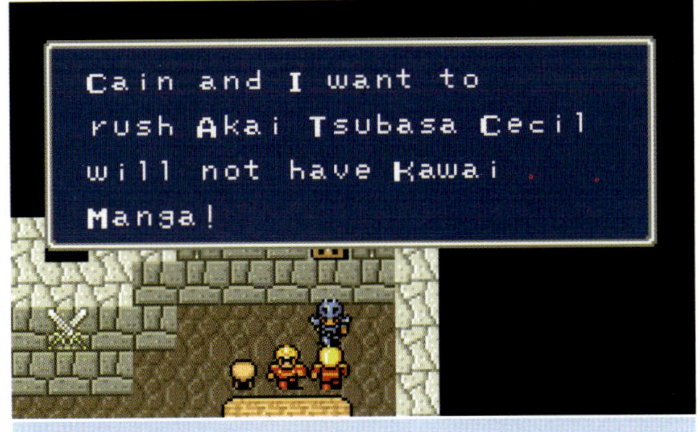

With Cain of the Dragon Knights and Cecil of the Red Wings, we have nothing to fear!

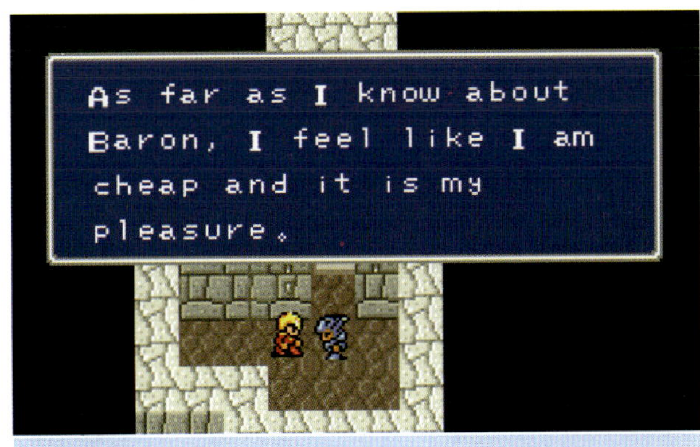

Treasures passed down for generations in Baron are stored beyond here, but entrance is forbidden.

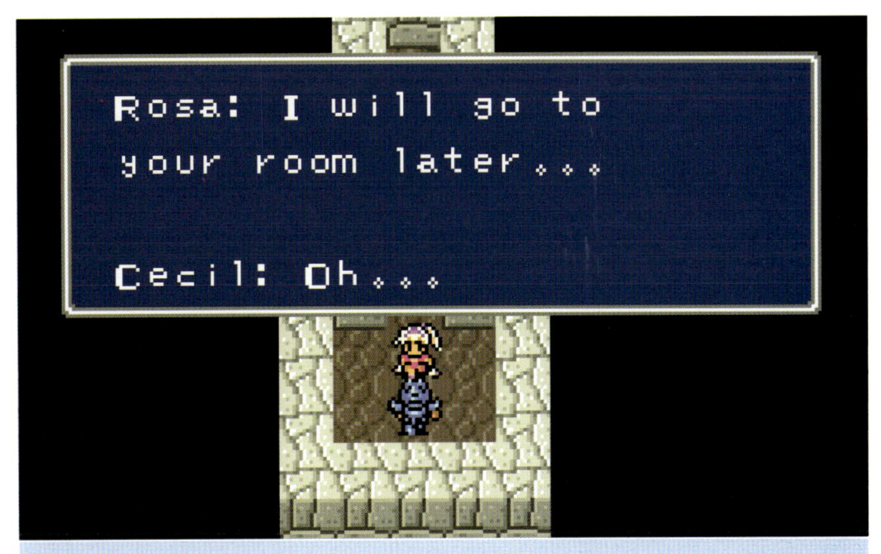

Rosa: I'll visit your room later...
Cecil: Okay.

▲ It's a Date

The machine translation of Cecil's reply to his girlfriend can be interpreted in different ways. Is it an "Oh..." filled with quiet excitement for what's to come, is it an "Oh..." filled with passive annoyance, or maybe something else? Human translations that look correct at first glance can often turn ambiguous on closer inspection. It seems the same holds true for machine translations.

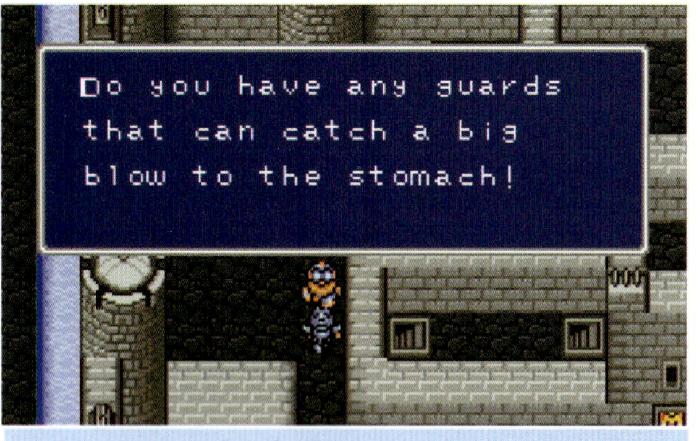

You're the only one who can lead the Red Wings!

Cecil: In Mysidia, I... I plundered a crystal from innocent people!

My daughter's been nagging me lately about how I never leave work!

◀ Pestered Parent

Despite the poor accuracy of Translation #2's text, it's sometimes possible to spot pieces of what the original script was trying to say.

In this instance, "so-and-so is loud" is basically the Japanese way of saying that someone is annoying or nags a lot. This is an example of how analyzing poor machine translations can provide some insight into how languages work. It might even make for interesting study material!

Story 

Before Cecil leaves for Mist Village, he explores the castle and meets with a few close friends. His girlfriend, Rosa, is particularly worried about Cecil's gloominess.

Mato Says 

You might notice that these game images have strange punctuation marks, like a Japanese-style circle period instead of standard English period. The original Japanese font doesn't include much English punctuation, so I improvised with what was available. I did go to the trouble of adding an apostrophe, though!

...they even became a tool for satisfying greed.

By His Majesty's order, you're not allowed into the castle until you've delivered the Bomb Ring to Mist Village!

▲ Village Not In

The Japanese word *mura* means "village", but there happens to be an unrelated word with the same spelling and pronunciation that means "unevenness" or "irregularity". Unfortunately, the machine assumed it was translating something scientific or industrial in nature. As a result, almost every instance of "village" in the Japanese script is translated as "unevenness" or "irregularity" in Translation #2.

Training soldiers in the way of the Dark Sword…
What is His Majesty thinking?!

Quick decision-making is essential in battle! But if it's too much for you to handle, you can lower the Battle Speed!

Story

Cecil and his friend Cain leave Baron Castle and head into town. There, they talk with the locals and prepare for their journey to Mist Village.

Reader Challenge

Quick! Think of another Japanese RPG that features laptop computers! *Persona* doesn't count.

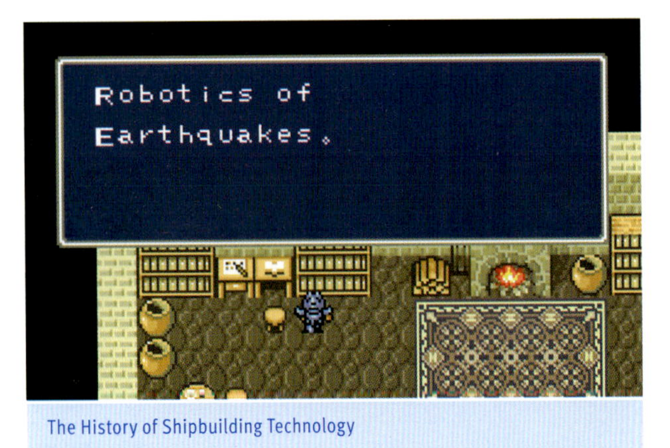

The History of Shipbuilding Technology

▲ Out of Time and Place

Even after multiple tests with the original Japanese line, I have no idea what caused the word "laptop" to appear here. This translation choice is certainly puzzling – laptop computers aren't something you'd normally expect in a medieval fantasy setting. But given that this Japanese line *is* already breaking the fourth wall, it's not as bizarre as I originally thought. It still remains incredibly inaccurate, however.

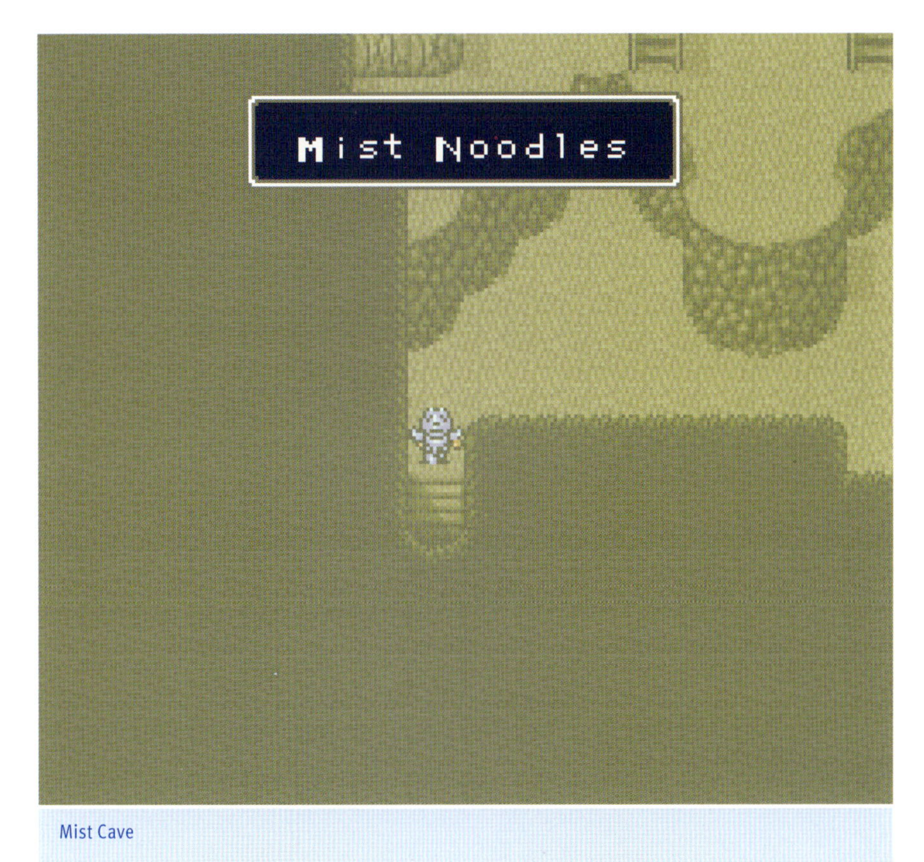

Mist Cave

Chance for a preemptive attack!

▲ Trouble with Teachers

The Japanese word *sensei* is well-known around the world as meaning something like "teacher" or "master". Unfortunately, *sensei* has many other meanings too, including "oath", "preemptive", and "despotism". Choosing between homophones – words with the same pronunciation but different meanings – is a challenge for any translator-in-training. As we see in this line, it's clearly a challenge for machine translators too.

Girl: Then you're the ones who killed my mom's dragon!

Cecil: We... had no idea doing so would kill your mom...

Story

Once Cecil and Cain reach Mist Village, the ring Cecil was ordered to deliver suddenly explodes. The village goes up in flames.

A young girl named Rydia blames Cain and Cecil for killing her mother. Enraged, she causes a massive earthquake that knocks everyone out.

Girl: Stay away from me!

▲ Regional Computer Dialects

There's no single type of English in the world – dialects and regional variations abound. So what kind of English can we expect from a machine translator?

Translation #2 features the occasional British English word, such as "mum" in this case, but it appears to lean more heavily toward American English overall. Interestingly, Translation #1 uses the British English word "bollocks" in this same scene. Is this just a coincidence, or is there something about this scene that appears especially British to machine translators?

Cecil: What a relief. She's not hurt. ... !!

General: Now, hand over the girl!
Cecil: I will not!

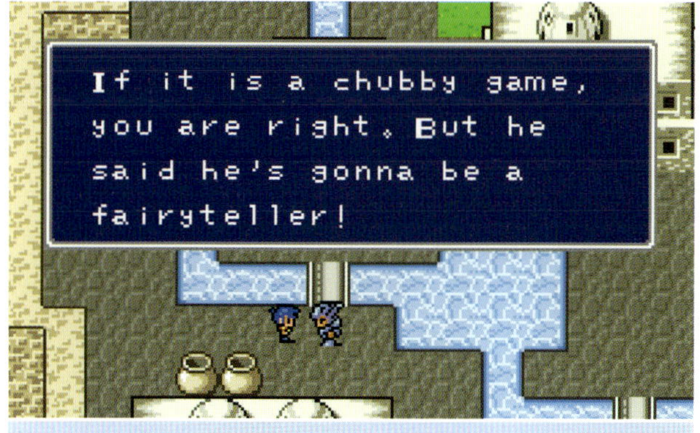

The underground waterway is to the northeast. But eight Water Snakes recently appeared there!

▲ Getting the Girl

In Japanese, the word *musume* usually means "daughter", but it can also just mean "girl" in certain situations. Without context – and the ability to even read context – the machine made the wrong choice when translating this line. I have no idea why Cecil is emitting a sharp ringing sound, though.

Story 

Cecil wakes up to find Cain missing and Rydia unconscious. He carries Rydia to the nearby Kaipo Village.

That night, Baron soldiers arrive to take Rydia away. Cecil defeats them and becomes Rydia's friend.

Mato Says 

Rydia's Japanese name can be spelled several different ways in English. Every official translation of *Final Fantasy IV* has called her "Rydia", but you'll notice that she's sometimes known as "Lydia" in the machine translations. This alternate name isn't a mistake – it's a perfectly valid translation.

For simplicity's sake, we'll refer to her as "Rydia" in this book.

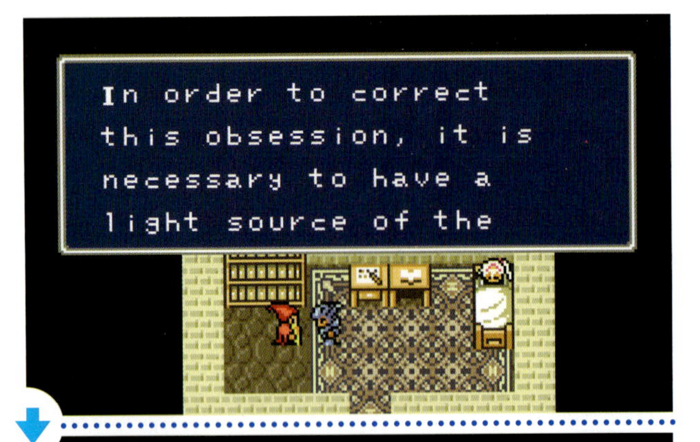

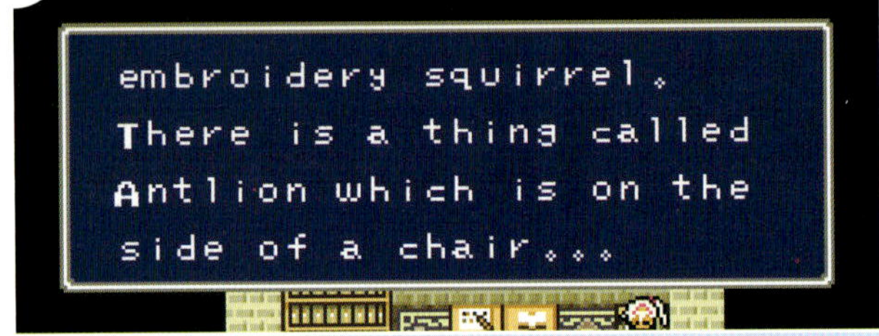

You'll need a mystical gem known as the Light of the Desert to treat her fever, but it's only found in a cave where the Antlion monster dwells.

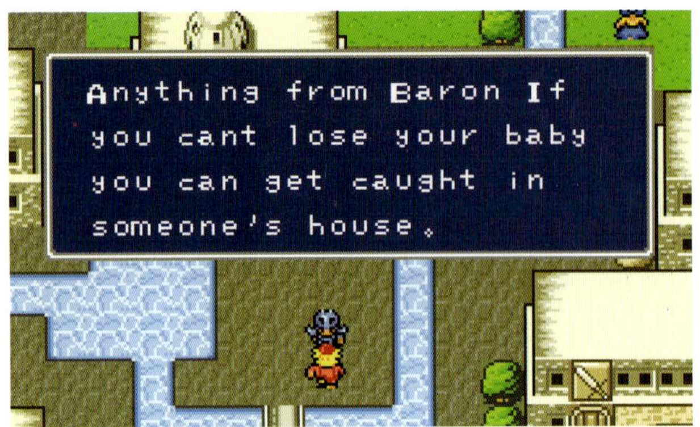

Apparently, a pretty girl from Baron fainted and was carried to someone's house.

If you're gonna take that little girl with you, I recommend you put her in the back row.

Story

Cecil learns that Rosa has fallen ill. The only cure is a rare item guarded by the Antlion.

Cecil and Rydia travel through a watery cavern. They meet Tellah, an old wizard who's headed to Damcyan in search of his daughter.

Cecil: My friend is suffering from a high fever in Kaipo.

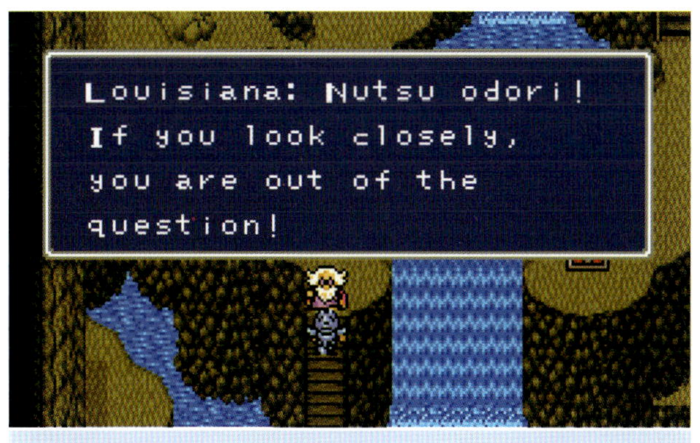

Old Man: Oh! You, there! On closer inspection, I see you're a wielder of the Dark Sword!

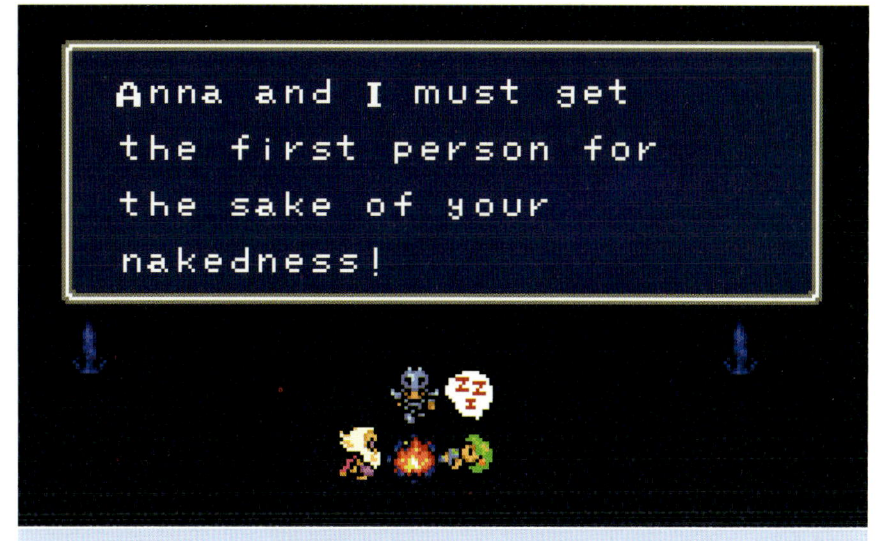

We must defeat it first, for Anna's sake and your friend's sake as well!

▲ Unexpected Nudity

It appears that the machine translator took the Japanese word *nakama* ("friend", "comrade") from this line and morphed it into *hadaka* ("naked") before translating it. This Japanese word-morphing is another pattern that's unique to Translation #2, as we'll continue to see.

Underground Lake

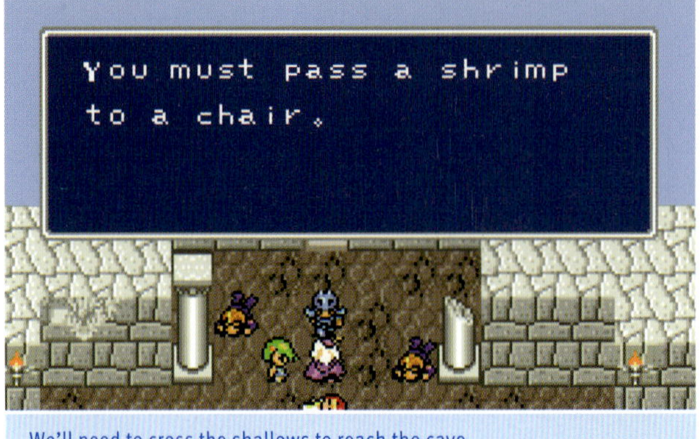

We'll need to cross the shallows to reach the cave.

Tellah: Just shut up!

▲ Crossing Words

In this translation, the word "shallows" became "shrimp" and the word "cave" became "chair". There's no logical reason for these translation choices, but they *do* follow a pattern we've seen before: the mistranslations share the same first letters as their proper translations. For some reason, the neural network responsible for Translation #2 tends to morph words before and after the translation process.

Story 

Just before Cecil, Rydia, and Tellah reach the kingdom of Damcyan, Baron airships bombard the castle.

Tellah's daughter dies in the attack. Tellah blames the prince of Damcyan, then leaves to get vengeance on his own.

Mato Says 

In every official translation of *Final Fantasy IV*, the prince of Damcyan is named "Edward". In the Japanese versions, though, he's known as "Gilbert".

For simplicity's sake, and because this project is based on the Japanese script, we'll stick with the "Gilbert" name in this book.

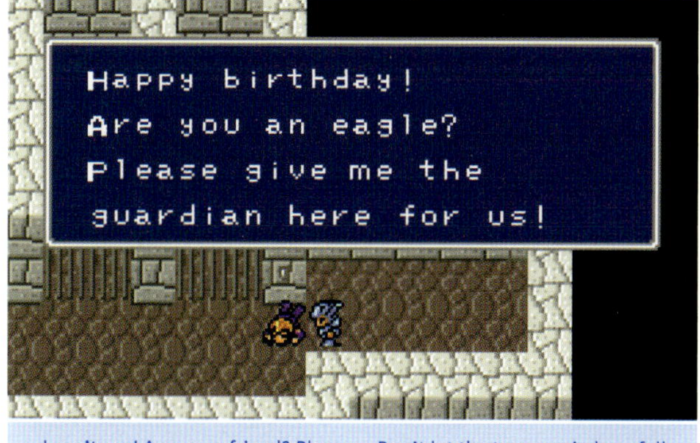

...I can't see! Are you a friend? Please... Don't let the treasure in here fall into enemy hands!

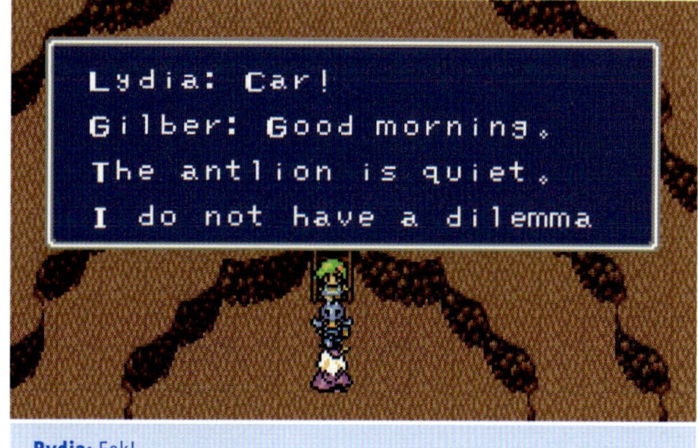

Rydia: Eek!
Gilbert: It's okay. The Antlion is tame. It doesn't harm people.

Story

Cecil, Rydia, and Gilbert enter the Antlion's lair, obtain its special curative item, and return to Rosa. Rosa is cured and joins the party.

Reader Challenge

The next time you're about to say "Oh my god!" or "WTF?!" after being surprised, shout "Car!" instead.

(Unshortened Text: "Japanese College College")
Antarctic Wind

▲ Car Talk

The Japanese shriek sound *kya* is very close to the Japanese pronunciation of the English word "car". Machine translation systems receive heavy use in the business and industrial sectors, so while I've never seen this particular mistranslation from a human translator before, it's certainly an understandable one coming from a machine.

The poor girl. She's caught a terrible fever and moans nothing but, "Cecil, Cecil..."

▲ Potty Mouth Madness

This kind old lady in the Japanese script uses the word "shit" twice in Translation #2. It seems that the machine translator took the word *uwagoto* ("to moan", "to talk deliriously") and replaced it with *tawagoto* ("nonsense").

I've seen the English word "shit" translated as *tawagoto* before, usually in sentences like "I don't wanna hear none of your shit." Unfortunately, "shit" has so many meanings and nuances in English that using the word improperly can give readers very different ideas than intended!

Oh, so you're the Cecil she was moaning about. Please take good care of her.

Cecil: Rosa! You need to wait here after all!

Rosa: I'll be fine. Besides, I'm a white mage. I won't slow you down!

Mato Says 

Rosa uses the phrase "all right" in this scene. This made me wonder if Google Translate disapproves of the alternate spelling "alright", so I checked all of Translation #2.

It turns out that "alright" appears 8 times throughout the translation, while "all right" appears 9 times. I guess Google Translate doesn't really care which spelling it uses!

▲ Addictively Simple

A basic Japanese 101 mistake caused the machine translator to mistake *wa* and *ha*. In simple terms, this led it to mistake *kimi wa matte* [...] ("you need to wait") for *kimi hamatte* [...] ("you need to be hooked on it").

I originally thought that the "must" in this translation was in the "I assume" sense, but it turns out it's "must" in the "you have to do it" sense. Accidentally adding extra interpretations like this is another pitfall that catches translators-in-training.

I have to go now... To become one with the Great Spirit...

Because of that Bomb Ring, Mist was...

Monk: I thank you for coming to my aid. I am Yang, the head monk of Fabul.

▲ Repeating Pattern

Once again, we see a mistranslated name that shares the same first letters as its proper translation. What's more, "Bomb Ring" was translated as "bamboo rope" earlier in the game, but here it's "bamboo rain". It seems that this letter substitution pattern isn't 100% predictable or consistent.

▲ Evolution of Language

Somehow the first instance of "monk" was translated properly, but the second instance transformed into "monkey" for no discernable reason. This type of indecisiveness is especially amusing in Translation #2 when multiple variations of the same word appear in the same scene.

Story 

Baron is planning to invade the kingdom of Fabul and take its crystal.

Cecil and his companions climb Mt. Hobs on their way to warn Fabul. On the mountain summit, they meet Yang, the head warrior monk of Fabul. He joins the party.

Reader Challenge 

Quick! We can't let the machines outsmart us! Think of one more word that starts with "monk"!

While we were training here on Mt. Hobs, monsters attacked. Everyone was wiped out but me.

▲ Tangled Text

Japanese grammar works very differently from English grammar, which means that long, descriptive sentences usually need to be untangled like a bundle of electrical cords before translation. The machine clearly failed to untangle this line, however, resulting in pure ridiculousness.

Gilbert: I'm the prince of Damcyan.

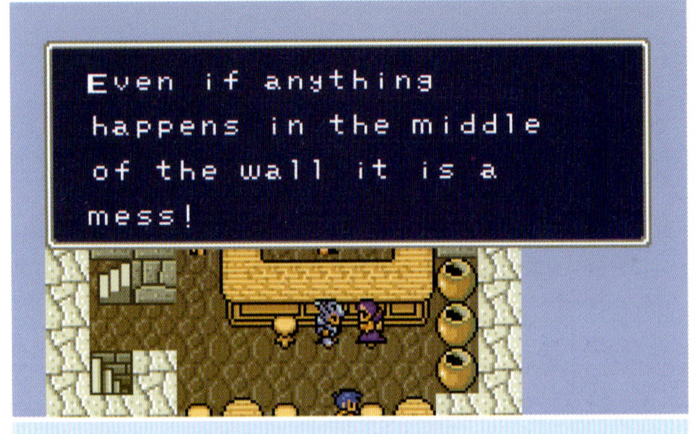

Even if something does happen, we're perfectly safe inside the castle!

Cecil: Ack!

Story

Cecil's party reaches Fabul and warns the king of the impending attack.

Cecil, Yang, and Gilbert help defend the castle from Baron's invaders.

▲ Getting Literal

The most interesting part of this mistranslation involves the phrase "inside the castle/wall", which works very similarly in Japanese and English. Taken literally, this phrase could refer to being inside the actual physical wall itself – hence this "middle of the wall" translation. Most of the time, though, it refers to being in the area enclosed by the wall. This is a great example of how important it is for a translator to capture the writer's intent rather than just the writer's words. At present, machines don't seem capable of this level of intuition.

They saved me from certain death... **Cecil:** Time is of the essence! **Rosa:** Please hurry!

Obtained 200 Gil

◂ Totally Blanking

Video game scripts often reuse single lines of text multiple times – sometimes hundreds of times – under different circumstances. It's also common for video games to play "fill-in-the-blank" with certain lines of text. This presents a unique challenge not usually seen in other forms of text translation. For example, in *Final Fantasy IV*, the line "Obtained __ Gil!" has a new number inserted into the blank after every battle.

Sentences with fill-in-the-blank content are problematic for machine translators, though – sometimes the blank will stay where it is, sometimes it'll move to a very different spot in the sentence, and sometimes the blank might get dropped entirely. During Translation #2's translation process, many of these blanks were corrupted in a way that would crash the game. So although my goal for this project was to have every step of the process 100% automated, I was forced to fix these specific problems by hand.

Cecil: At this rate, we'll be destroyed!
Yang: Let's retreat into the castle for now!

Door unlocked!

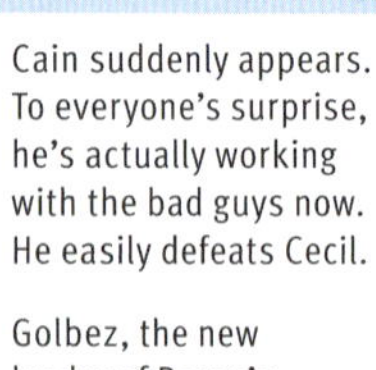

Story

Cain suddenly appears. To everyone's surprise, he's actually working with the bad guys now. He easily defeats Cecil.

Golbez, the new leader of Baron's military, makes his first appearance. He defeats the rest of the heroes with a single magic spell. He then takes Fabul's crystal and kidnaps Rosa.

Tales of Vesperia
(Xbox 360)

Confusing "key" and "lock" is a classic Japanese-to-English translation mistake that persists to this day.

▲ Keywords

Many aspiring translators get tripped up by the Japanese word *kagi*, which can mean both "key" and "lock". The intended meaning is usually clear from context – it's very rare that you'd ever ask someone to insert a lock into a door, for example.

As we've already seen, machines have difficulty recognizing contextual clues. Consequently, this generic key/lock mistake is a strong (but not 100% guaranteed) indicator of a Japanese-to-English machine translation.

His Majesty was injured in the attack and is now resting in his bedchamber.

I *am* your wife, you know! Some Baron soldiers showed up here, but I bashed 'em with my frying pan.

Story

Cecil and his companions decide to sneak back into Baron via boat.

After the king of Fabul hears about the current situation, he gladly lends the heroes a ship.

If we had just one airship, we'd actually be able to fight back!

▲ Wife Material

By itself, the Japanese word *tsuma* usually means "wife". Unfortunately, a completely unrelated word with the same pronunciation appears in phrases relating to fingertips. Confusing these two versions of *tsuma* would be like mixing up "I" and "eye" in English – it's an understandable mistake but surprising to actually see.

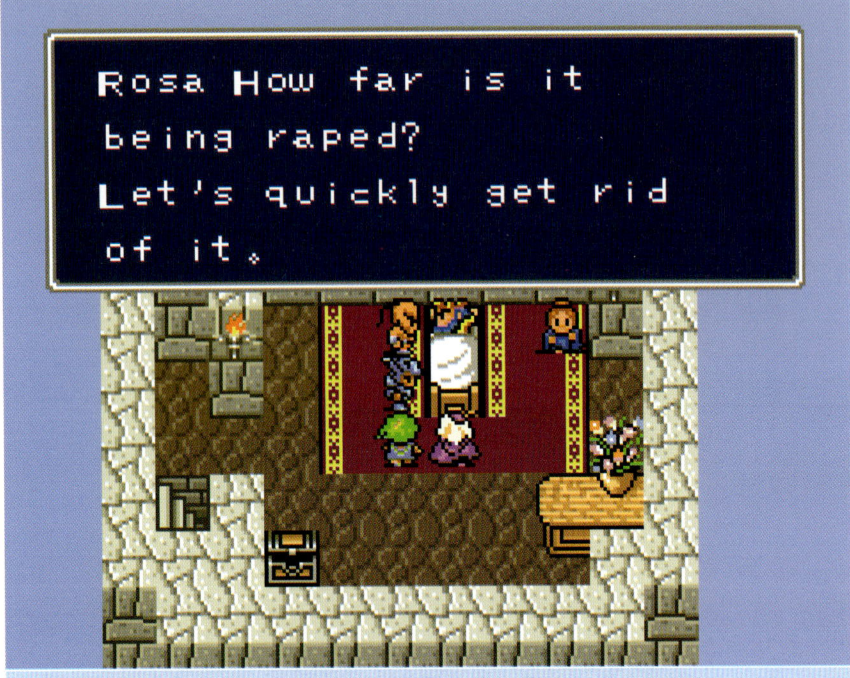

They even abducted Lady Rosa... I will ready a ship at once.

◄ Translation and Situational Awareness

There comes a time in every translator's journey when, while in the middle of translating a line, something seems off. "This doesn't seem right. Am I mistranslating this?" is what most human translators might think, but machine translation systems still lack the ability to stop and realize when a translation has taken a wrong turn.

Additionally, humans can usually identify when a translated line of text has turned into unintentionally offensive content. It's clear that the machine translator was unable to make that determination with this particular line. In short, current machine translators not only have trouble with reading and translating, but are also blind to the qualities of their translations.

Out of curiosity, I looked further into how this mistranslation might've happened. I experimented with the Japanese phrase in question, and even after several dozen tests I could find no logical reason for this translation choice. Similar mistranslations happen elsewhere in the script, but for equally mysterious reasons.

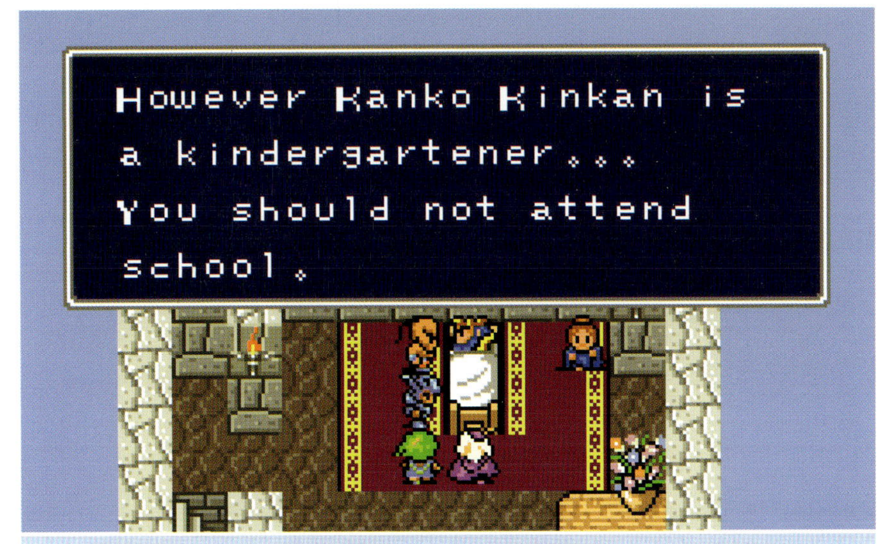

Even so, darkness is at the heart of the Dark Sword arts. Your blade will have no effect against true evil.

▲ Royal Schooling

This mistranslation is full of issues, the most perplexing of which is "Kanko Kinkan". Where did this phrase come from? What does it mean? After running many tests on the Japanese sentences, I still remain baffled. Nothing like it appears in the Japanese text – it's as if the machine randomly created new words out of thin air.

Defeat Golbez, without fail.

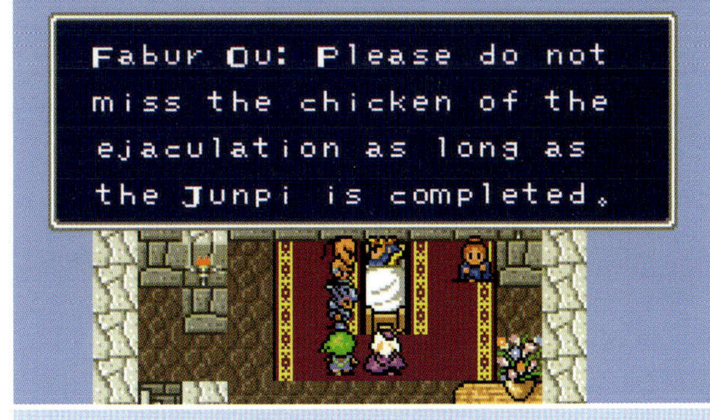

King of Fabul: Once you are ready, go to the eastern pier.

We will do our best to defend the castle.

It's the Fat Chocobo!

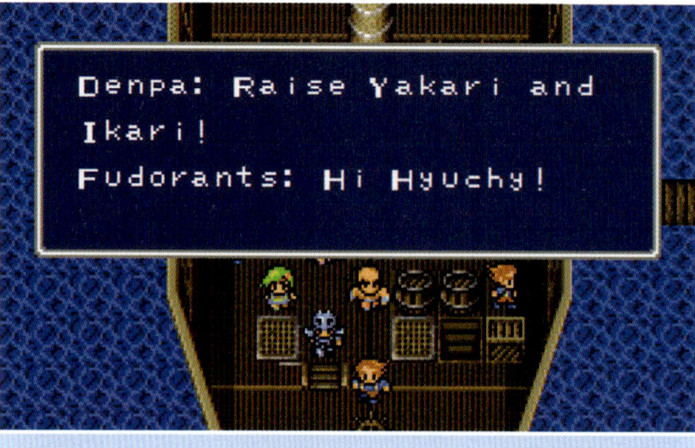

Captain: Raise the anchor, men!
Sailors: Hyaho!

▲ Jumping Ship

This line is a perfect example of how the cutting-edge translation system sometimes gives up and invents new Japanese-sounding words that aren't actual words at all. It's what I'd expect if a lazy human translator got tired and decided, "Whatever, I don't understand what they're saying here so I'll just make up something. Nobody will care anyway." Except in this case it's a machine being lazy.

Story 

The party boards a ship bound for Baron, but a sea monster known as Leviathan attacks. Everyone is thrown overboard.

Mato Says 

In the *Final Fantasy* series, "Chocobos" are friendly birds that help the player. Some Chocobos act as fast transportation, some can heal your party, and others can store your excess items.

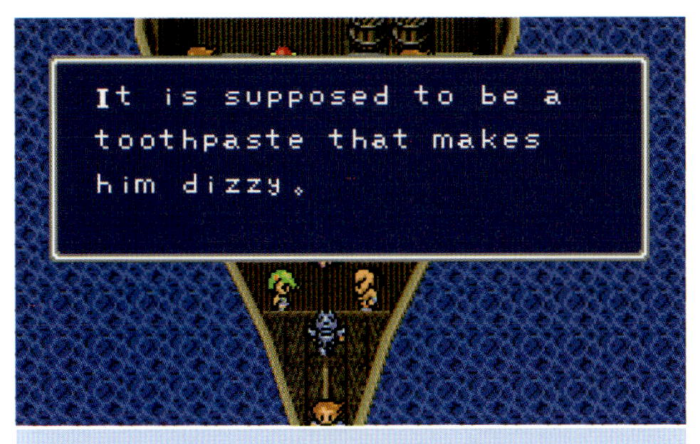

He knows all about airships and will definitely help us out.

(Unshortened Text: "I was lucky at the tea ceremony!")
Fell into an incapacitated state!

Story

Cecil wakes up all alone on an unfamiliar shore. In the distance is Mysidia, the town that Cecil pillaged at the beginning of the game. Naturally, none of the citizens are happy to see him again.

Yang: I hope that the dumplings are good.

Yang: I hope that he's all right.

Armor Shop

Shaddup! *hic* Ya don't scare me, Dark Knight!

▲ Translation Inebriation

Drunk talk can be difficult to understand in any language, and even more challenging to translate. In this instance, a drunken wizard's dialogue proved too difficult for the machine to process, so it inserted nonsensical Japanese-esque words as a last resort instead. Now the line comes across as a completely different type of incoherent rambling.

Our friends! Return our friends!

▲ Give It Back

This baffling mistranslation is also one of my favorites. While trying to figure out how this mistranslation occurred, other phrases like "Return the bastard!" and "Return the baby!" popped up as alternatives. Based on the patterns we've seen so far, I'm guessing the machine is changing the word "brethren" or "brothers-in-arms" into completely unrelated words that start with "b".

Back then, I lacked the courage to oppose the king's orders.

Story

Cecil asks Mysidia's elder for help. The elder says that Cecil must first visit Mt. Ordeals to shed his dark side and become a Paladin – a warrior of light.

Two wizards in training, Palom and Porom, are told to accompany Cecil.

▲ Receding Hare Line

While examining how this rabbit-related mistranslation occurred, I encountered alternate machine translations such as "There was no remedy for the miscarriage of a mother." and "There was no remedy for the male nude."

This translation is utter nonsense, but the "bunny" part does make some sense from a translation perspective. The machine presumably saw the Japanese character *u* in the sentence, mistakenly decided it was the name "U" (the Rabbit of the Chinese zodiac), and translated it as "bunny". It'd be like if you took this sentence you're reading now, saw the capital I at the beginning, and translated the entire word as "me".

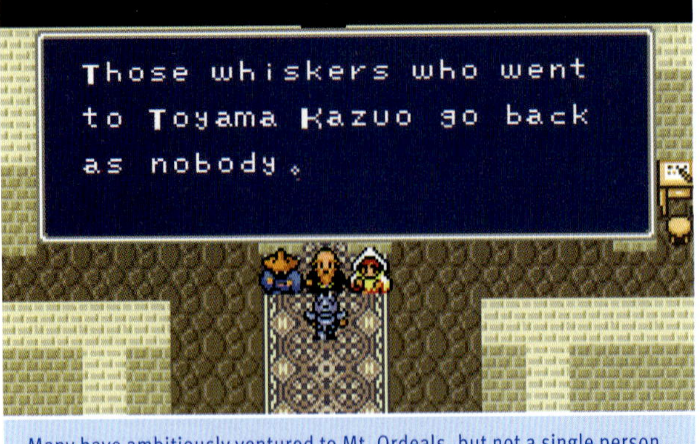

Many have ambitiously ventured to Mt. Ordeals, but not a single person has returned.

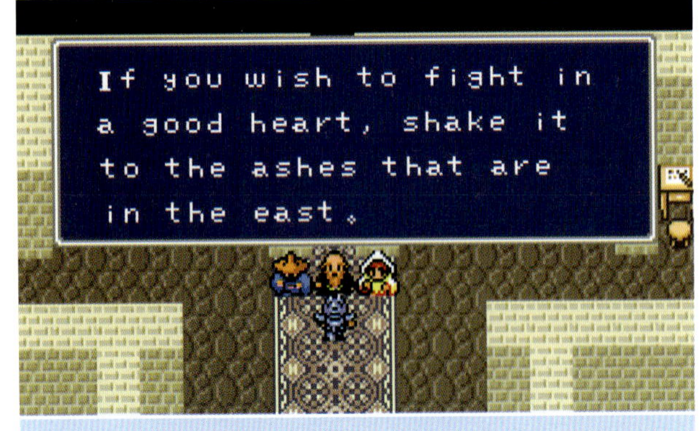

If you wish to fight with a heart of good, then go to Mt. Ordeals to the east.

◀ Surprise Shout-Out

This translation includes the name of a random Japanese man, but why? First, we'll need to take a look at this key part of the Japanese line:

> しれんのやまへいったものは　かずおおい
> *shiren no yama e itta mono wa kazu ōi*

It seems that the machine saw the "kazu" and "ōi" at the end, partially combined them, and decided that the line was about someone named Kazuo. The "Toyama" part must've come from the "yama" that's also in there, but it's unclear how it transformed into "Toyama". It's a truly bizarre translation choice all around. In any case, here's hoping one of the Toyama Kazuos out there discovers this book!

On a side note, this "Toyama Kazuo" name is written in Japanese name order, which goes family name first, given name second – so outside of Japan he'd actually go by "Kazuo Toyama". Basically, the machine inserted a Japanese name into this English translation, but inexplicably left it in Japanese name order.

Mato Says 

Mysidia's town elder is simply called *chōrō* ("Elder") in the Japanese script, but he goes by many different names throughout this machine translation, including "Chiucho" and "Welcome". Out of all the names he received, I'm particularly fond of "Chiropractus".

Reader Challenge 

If you know a Kazuo Toyama in real life, let him know about this book. Otherwise, the closest object to your right is now named Kazuo Toyama. Tell him about this book.

▸ Eagle Mania

As we've seen, the Japanese language is filled with confusing homonyms and homophones. A great example of this is the word *washi*.

In some cases, particularly in entertainment, *washi* is a pronoun that elderly people use to mean "I" or "me". *Final Fantasy IV* has many elderly characters who use *washi* in this way, in fact. Unfortunately, *washi* can also mean "eagle" in other situations. Context usually makes it obvious which meaning is intended, though. For example, if you point at an eagle and say *washi*, the listener will naturally assume you're talking about the bird and not about yourself like an old person.

Because machines still have trouble recognizing context, Translation #2 is teeming with old people who talk about eagles non-stop. Their obsession with eagles is so strong that eagles became the honorary mascot for Translation #2.

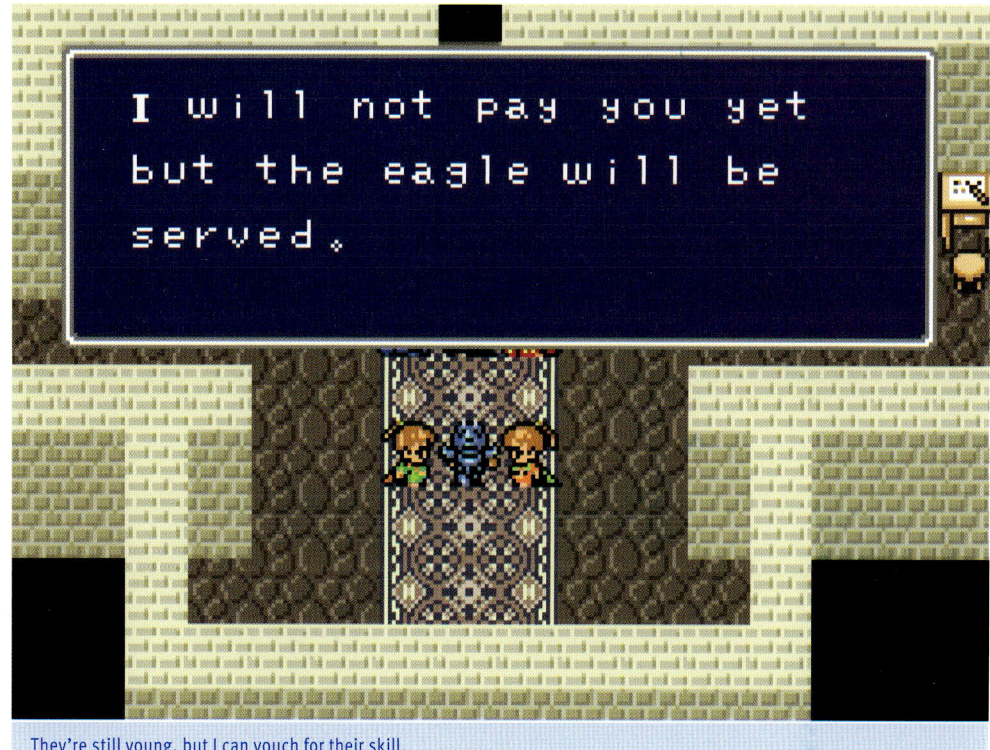

They're still young, but I can vouch for their skill.

Your fate is in your hands. All I can do now is pray...

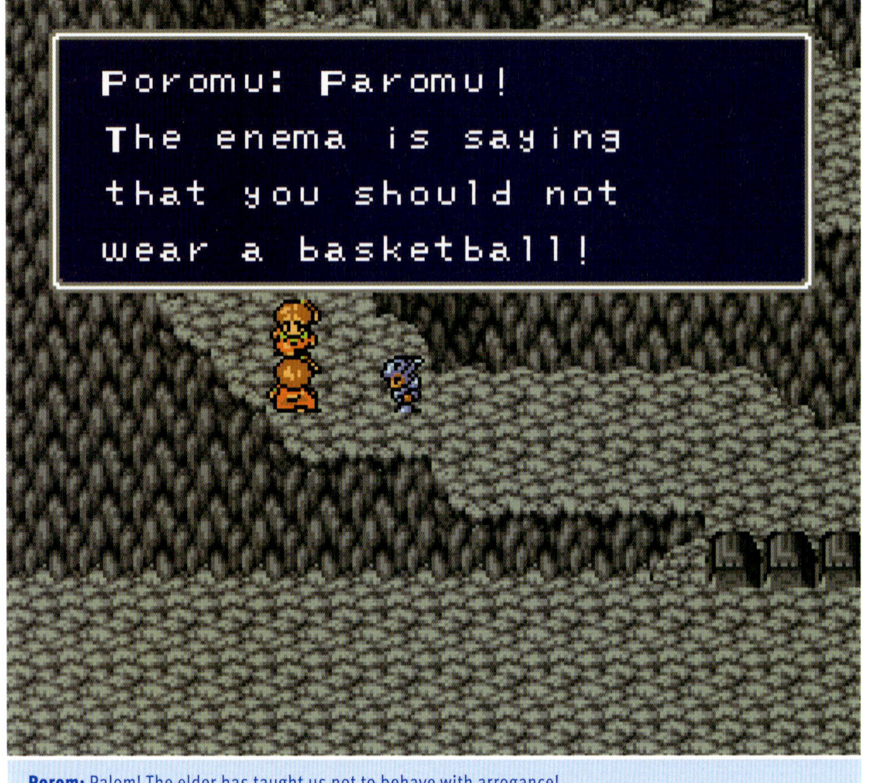

Porom: Palom! The elder has taught us not to behave with arrogance!

◀ Sage Advice

This is one of *the* finest mistranslations I've ever seen. There's no sense to it on the surface, but a look at the original Japanese text clears up some of the mystery.

First, the word *chōrō* means "elder" in situations like this. It turns out that there's a separate word with the same pronunciation that means "enteric fistula", which is an abnormal connection between the gastrointestinal tract and other organs. I had never heard of this second meaning of *chōrō* before, so this came as a big surprise. I can only assume that this anatomical term about intestinal stuff is how the machine translator reached the word "enema".

Next, the word *takaburu* ("to behave conceitedly") was incorrectly identified as the word *kaburu* ("to wear on the head"). This mistake is a little understandable, and I could see an inexperienced human translator mistaking *takaburu* for something else.

After many experiments with the original Japanese line, I still have no idea how a basketball appeared in this mistranslation. Instead, I encountered alternate translations such as "You said hero not to be covered with chopsticks" and "The hedge says that you should not bear up".

Story 

Cecil and his wizard companions reach Mt. Ordeals. They meet Tellah, who seeks the ultimate, forbidden magic spell. Tellah joins the party.

Meanwhile, Golbez sends the Fiend of Earth to stop Cecil from becoming a Paladin.

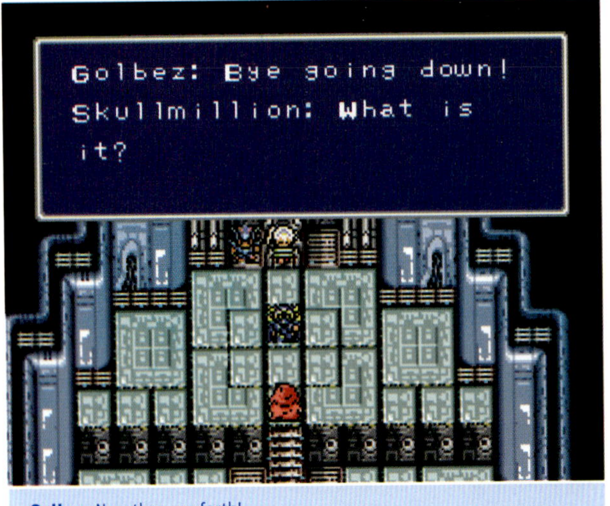

Golbez: Now then, go forth!
Scarmiglione: As you wish.

▲ A Million Names

The name "Scarmiglione" was translated many different ways in Translation #2, but never correctly. Of the many incorrect translations the name received, "Skull Million" turned out to be surprisingly appropriate for this undead monster lord.

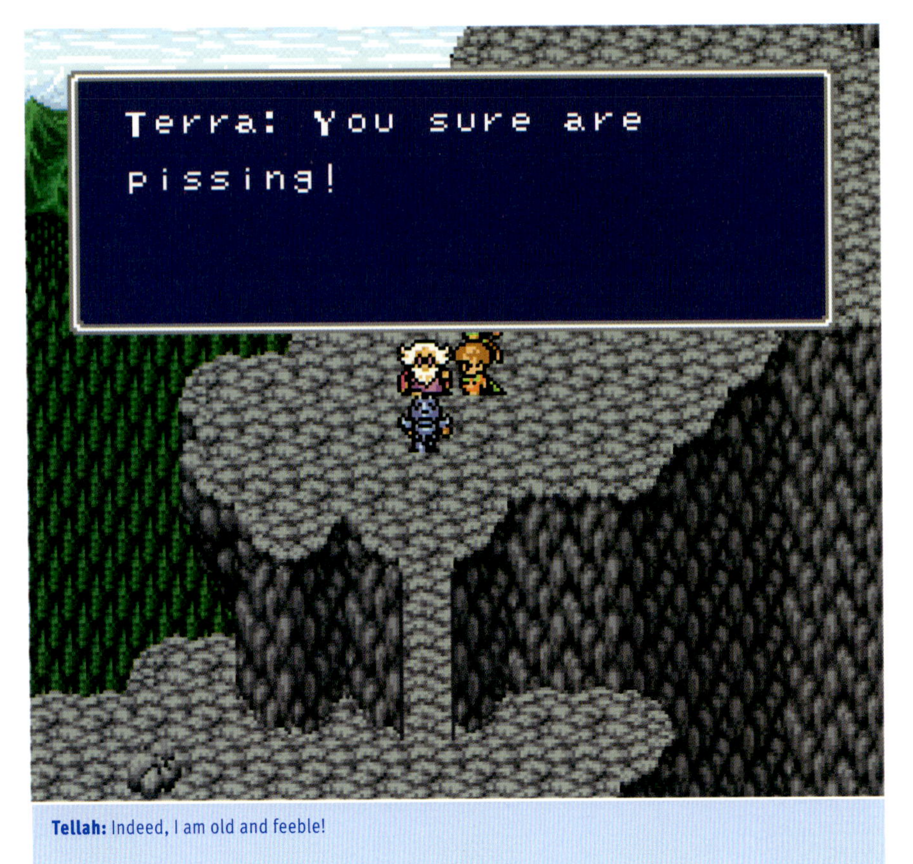

Tellah: Indeed, I am old and feeble!

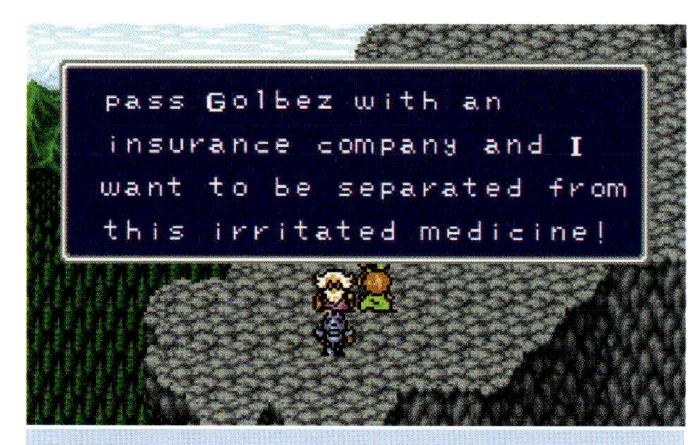

I'm told Golbez can't be defeated with the Dark Sword arts... I want to be free of this accursed Dark Sword!

▲ Life Policy

It's fairly common knowledge that machine translations aren't 100% reliable and that they should rarely be used as-is. From this understanding, though, comes the misconception that it's safe to take a machine translation, rewrite it based on what the translation *seems* to say, and then present the rewrite as a proper translation. This practice of rephrasing machine translations may be cheap and quick, but it also carries great risk. For example, just imagine someone trying to rephrase this crazy line from Translation #2 – any rewrite would still be nonsense!

Poisoned!

My son...
Cecil: Son?! Who are you?

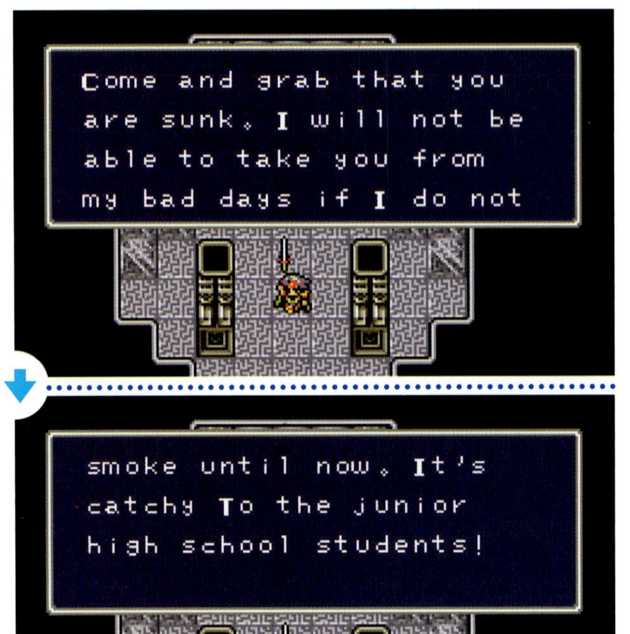

Now, break ways with your blood-soaked past. You must conquer your former self to receive this holy power. Overcome the Dark Knight in you!

▲ Hearing Things

In this translation, the Japanese word *musuko* ("son") was replaced with the English name "Susan". There's no logical reason for this change, but it does follow the First Letter Pattern that we've seen before. This emphasis on first letters isn't how translation works, though, and no human translator would ever make this same mistake. In a way, this goes against the general notion of machines being more logical than people!

Story 

The heroes defeat the Fiend of Earth after a difficult battle. They soon arrive at a mysterious shrine filled with mirrors.

Cecil is forced to face his dark side in battle before he can become a Paladin.

Reader Challenge 

Quick! What were two other things that were catchy when you were a junior high student?

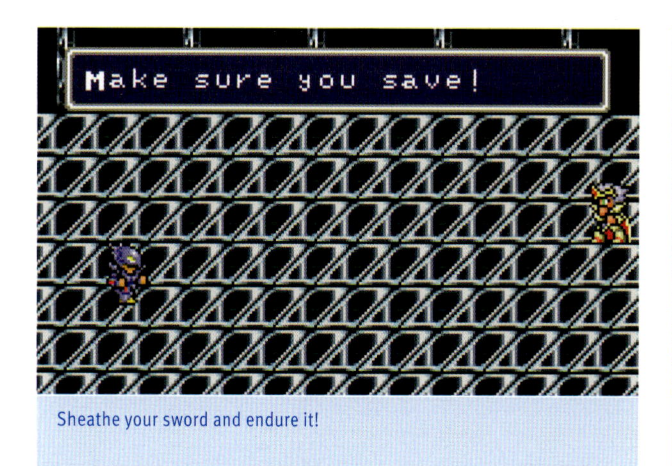

Sheathe your sword and endure it!

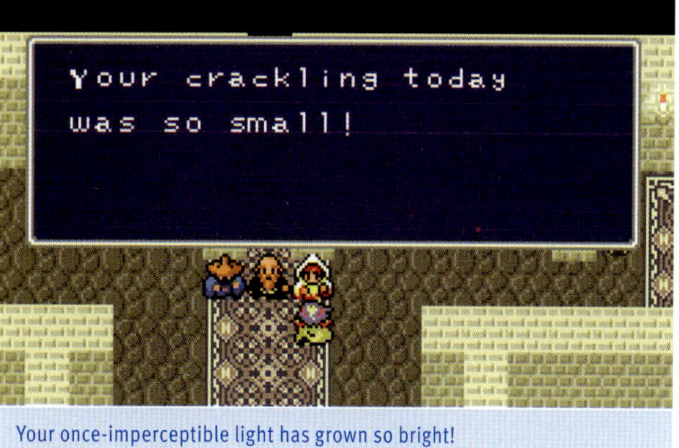

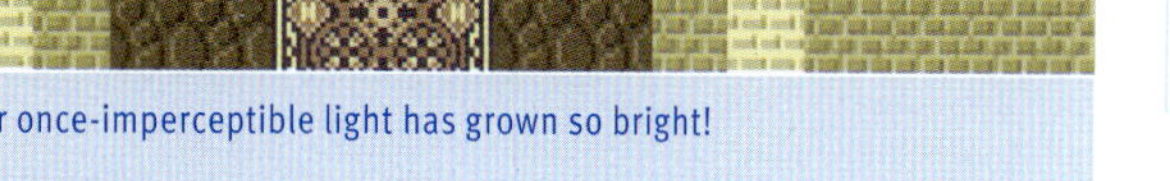

Your once-imperceptible light has grown so bright!

Story

Cecil becomes a Paladin with a heart of light. The people of Mysidia are pleasantly surprised, including the elder.

The elder shares a legend that speaks of Cecil's transformation.

Palom and Porom remain by Cecil's side, even though their task is over.

Never forget the Paladin spirit.

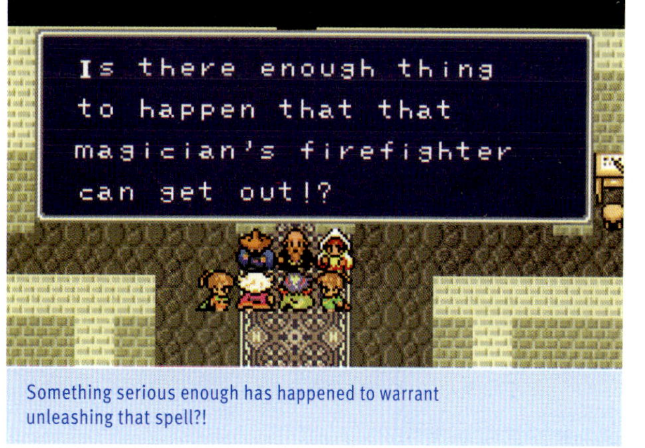

Something serious enough has happened to warrant unleashing that spell?!

Things that are
born from the mouth of
the crown
Taikaku Mai Ra
Bring darkness further
after sleeping with
darkness and light.
The next is surrounded

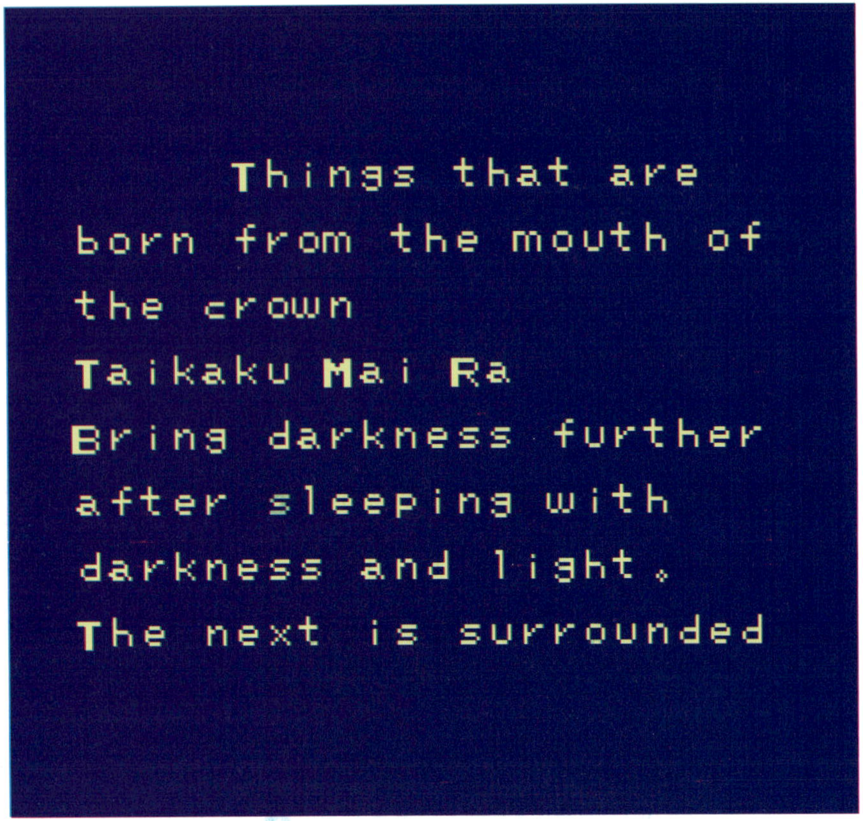

Machine Translation	Intended Translation
Things that are born from the mouth of the crown	One born from the mouth of a dragon
Taikaku Mai Ra	shall soar high into the heavens,
Bring darkness further after sleeping with darkness and light.	and, hoisting the dark and light, bring forth a greater promise unto the land of slumber.
The next is surrounded by Hikisaki Hikari	The moon, veiled in everlasting light,
Hajime Tachikawa gives a big Megumi Jihi.	bestows a great bounty and mercy to Mother Earth.

▲ Legendary Importance

An ancient legend lies at the heart of *Final Fantasy IV*'s story. The Japanese version of the legend is written in a slightly archaic way that's quite different from everyday speech. The machine translator had difficulty parsing this non-standard text style. What's more, the machine translation was significantly longer than the game's programmers had originally intended. As a result, the final two lines of the translation never get displayed.

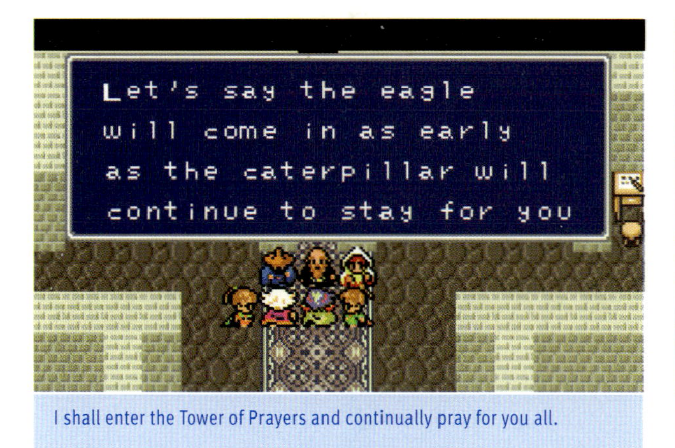

I shall enter the Tower of Prayers and continually pray for you all.

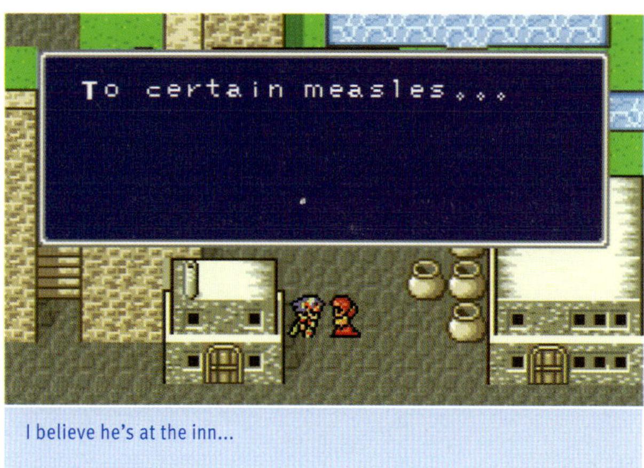

I believe he's at the inn...

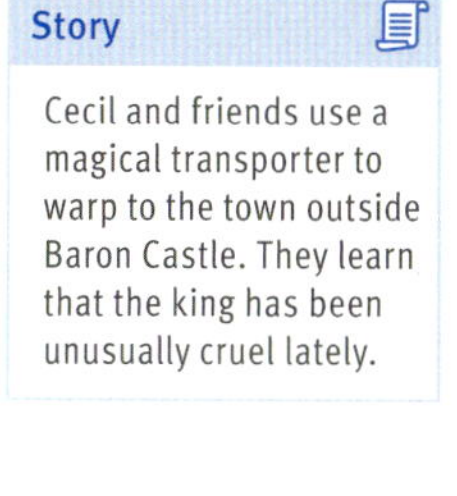

Story

Cecil and friends use a magical transporter to warp to the town outside Baron Castle. They learn that the king has been unusually cruel lately.

I wonder if that Dark Knight guy really did die. I liked him.

▲ Translatable Diseases

It appears that the machine translator took the Japanese word *tashika* ("I believe", "If my memory is correct") from this line and changed it to *hashika* ("measles"). Tashika is a very common, everyday word that appears all the time. In contrast, *hashika* is rarely a topic of normal conversation in Japan. The machine behind Translation #2 probably doesn't take "commonness" into account at all, but that doesn't explain why it changes words in the first place.

The door of the building on the west side of town? It connects to an underground waterway that leads into the castle. It's sealed now, though.

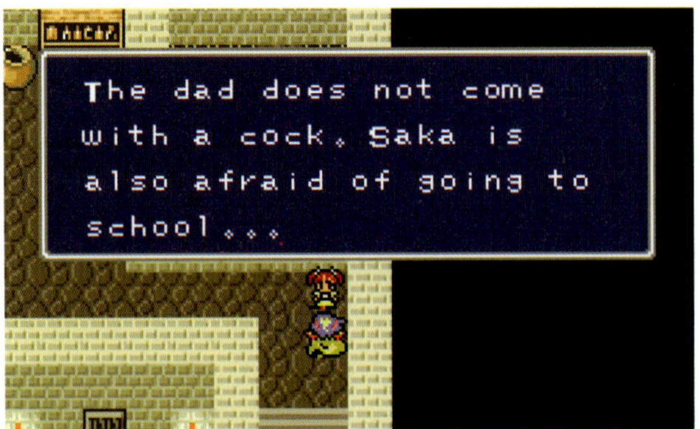

My dad says we aren't getting any customers anymore. And there are scary soldiers at the pub, too...

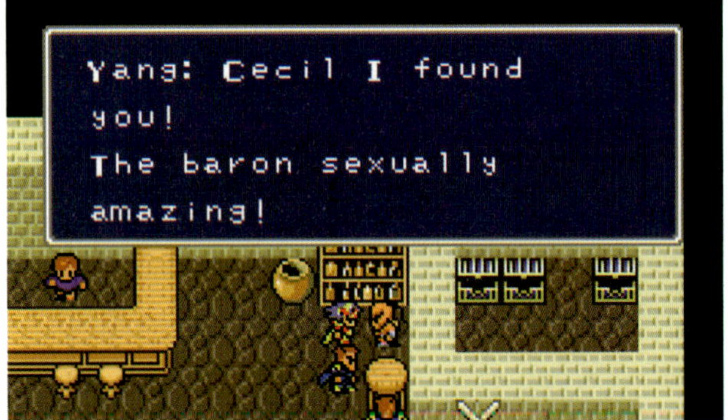

Yang: I finally found you, Cecil! You traitorous dog opposed to the King of Baron!

Story

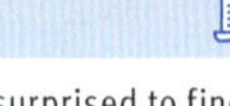

Cecil is surprised to find Yang in the local pub. Unfortunately, Yang has been brainwashed to work for the bad guys. He attacks Cecil.

After a tough battle, Yang breaks free of his brainwashing and rejoins Cecil's party.

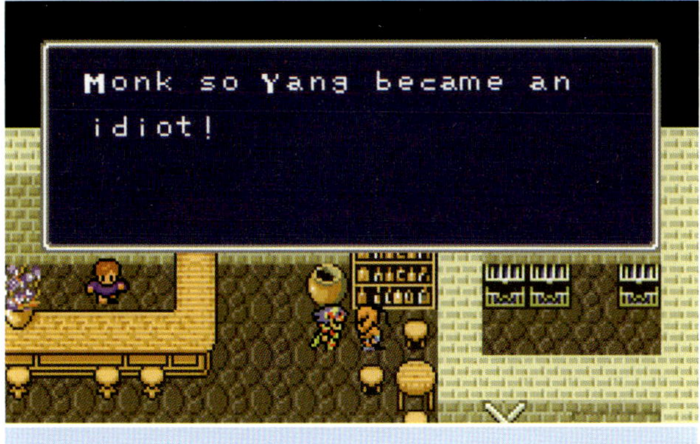

The monk Yang joined the party!

Tellah: But we cannot simply waltz right up and go inside...

Story

Cecil wants to meet with the king of Baron, but guards block the castle entrance. The heroes sneak into the castle via an underground tunnel instead.

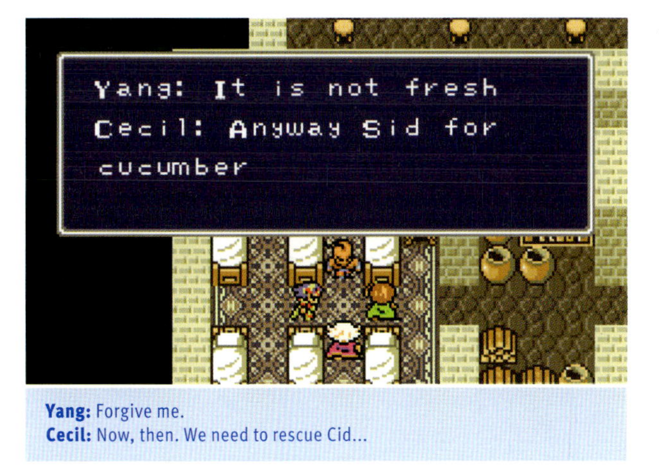

Yang: Forgive me.
Cecil: Now, then. We need to rescue Cid...

▲ Mashed Mistranslations

In this line, the phrase *ja ga* ("but..." as said by an elderly person) was mistaken for the word *jagaimo* ("potato"). Because this is such a basic misunderstanding, the potato mistranslation is relatively consistent throughout the script. As a result, elderly characters in Translation #2 have an obsession with potatoes that almost matches their fascination with eagles.

What do you want? Do you have a pass?

◀ Travel Agency

This machine translation includes a reference to Fukuoka, an actual city and prefecture in southern Japan. Fukuoka is known for its temperate climate and strong tourism industry, so this mistranslation is surprisingly in tune with the region's reputation.

Of course, how a character in a fantasy video game knows about real-life Japan is another question entirely. There's no logical equivalent for "Fukuoka" in the Japanese text, so the machine must've seen something beyond human comprehension.

Incidentally, this reference to Japanese geography isn't an isolated incident – more locations get mentioned too. For whatever reason, Translation #2 offers players a crash course in Japanese vacation destinations!

Beigan: Oh, you're unharmed!
Cecil: Beigan! Don't tell me you're also...

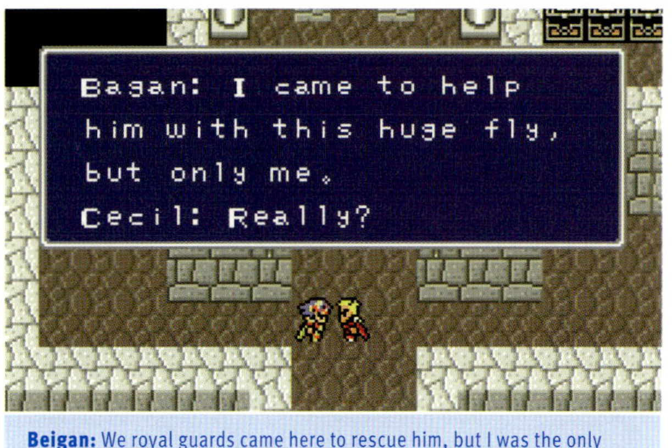

Beigan: We royal guards came here to rescue him, but I was the only one who survived. **Cecil:** I see.

Story

Cecil and friends explore the castle, but it's eerily deserted. Eventually, they meet a royal guard who transforms into a monster.

The heroes defeat the monster and head to the throne room. There, they find the king.

▲ Royal Roll Call

The neural network behind Translation #2 tends to be wishy-washy when it comes to translating unique names. For example, this royal guard – who is normally known as "Beigan" – goes by half a dozen names in Translation #2, including "Bagan", "Vegan", "BEIGAN", and the illustrious "Big gun".

As we've seen, this issue with indecisiveness and inconsistency is a constant problem throughout Translation #2's text. It's one of Translation #2's most unique characteristics, in fact.

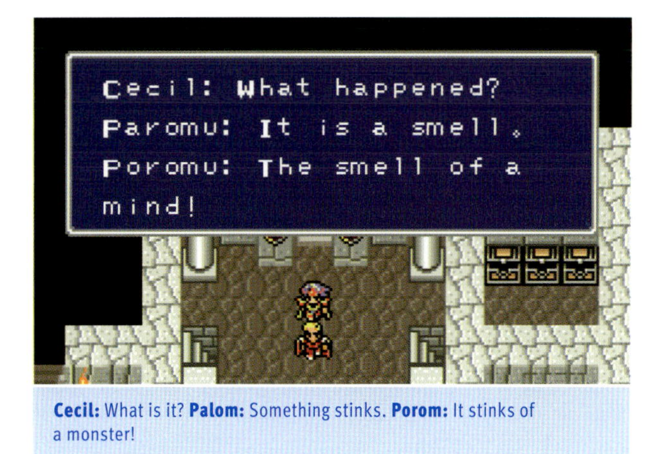

Cecil: What is it? **Palom:** Something stinks. **Porom:** It stinks of a monster!

Entangle

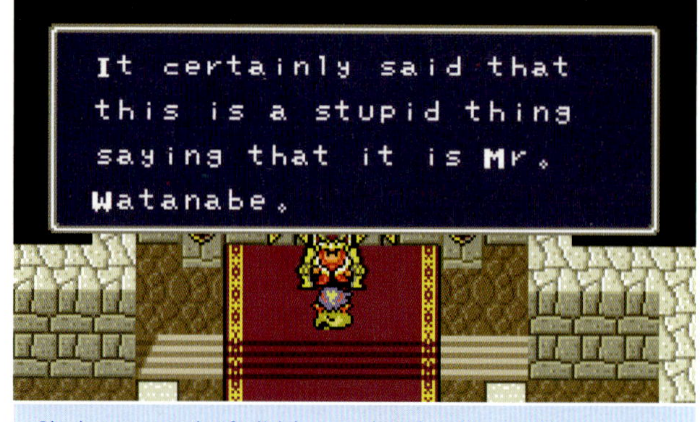

Oh, do you mean that foolish human who refused to give this kingdom to me?

▲ What an Abe

The phrase *watasan* ("I shan't hand it over") appears in this Japanese text. Unfortunately, it seems that the machine translator had trouble identifying this unusual wording and mistook it for *Wata-san*, the equivalent to something like "Mr. Wata" or "Ms. Wata". From there, the machine further assumed that "Wata" was short for "Watanabe", a common Japanese family name.

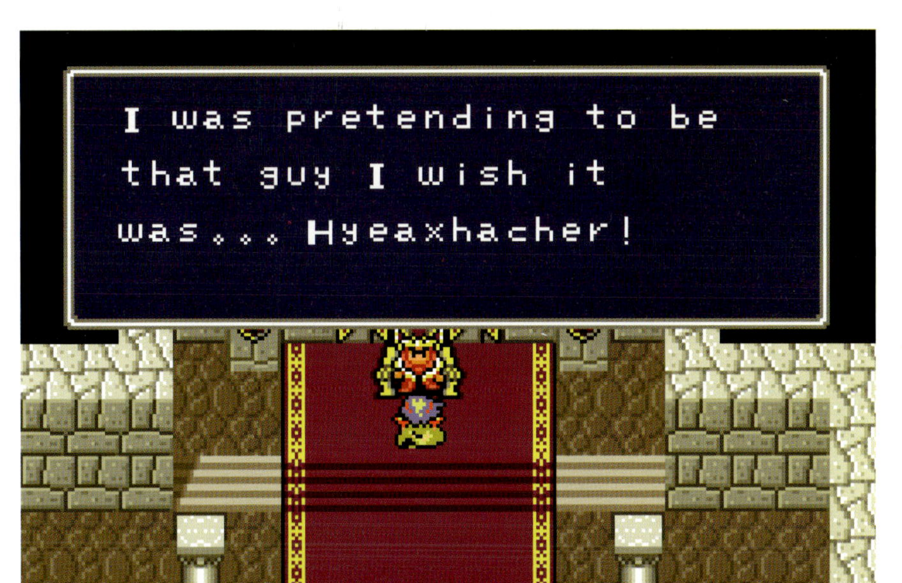

Oh, that's right, and I was pretending to be him all this time... Hyahyahya!

Dorry from the *One Piece* series demonstrates his unique "gegyagyagya" laugh

◀ Laughter Transformed

Japanese entertainment features an extremely wide variety of cackling and laughter, so it's common for characters to have their own unique laughing styles. Japanese laughter is a constant challenge for human translators, so I was eager to see how a fancy A.I. machine translator might handle it.

In this instance, the machine failed to recognize the villain's "hyahyahya" laughter in this scene. Instead, it seems it altered the laugh little by little in an attempt to find a similarly pronounced word. Then, after vainly trying to find a match, the machine gave up and went with its final guess.

He's on the defensive and hiding in his shell!

You filthy fake King of Baron!

Story

The king reveals that he's actually the Fiend of Water. He had killed the real king and taken his place much earlier.

After the heroes defeat the impostor, the airship mechanic Cid arrives and joins the party.

▲ Gotta Snap 'Em All

This is one of my favorite lines in Translation #2 – partly because it sounds so silly, but mostly because it makes no sense from a technical translation standpoint. What does "Jigglyphoto" even mean? Is it a Pokémon? A sexy photograph? Unfortunately, there's no way to really ask machine translators about their thought processes, so this question will forever remain unanswered.

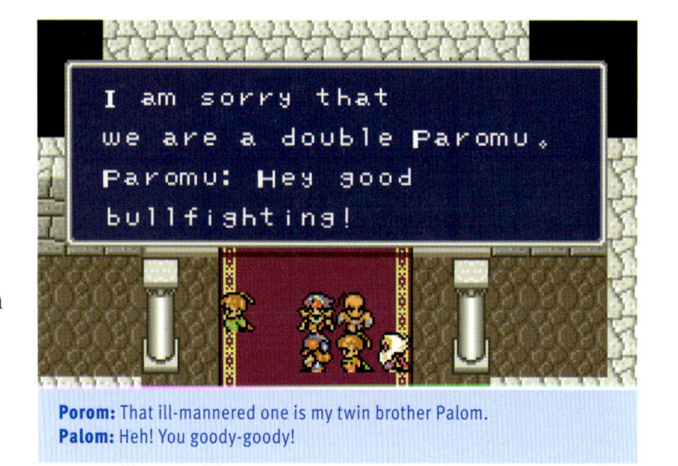

Porom: That ill-mannered one is my twin brother Palom.
Palom: Heh! You goody-goody!

Yang: It's nice to meet you. It's dangerous here. We must hurry.

◀ Artificial Unintelligence

As we've seen, the machine behind Translation #2 has no qualms with giving bizarre, offensive mistranslations.

I never thought about it before, but perhaps there'll come a time when machine translators take the vague idea of "appropriateness" into account. Not everyone shares the same values, though, so it would certainly be a difficult concept to implement. Who knows, maybe the machine translation systems of the future will have other settings like "morality", "creativity", and "audience awareness" that can be tweaked for specific target audiences.

In any case, I wasn't able to pinpoint what phrase in the source text led to this specific translation, but I did discover that simply changing the punctuation would produce alternate phrases like "I will congratulate you on the rack" and "I am congratulated by the snack".

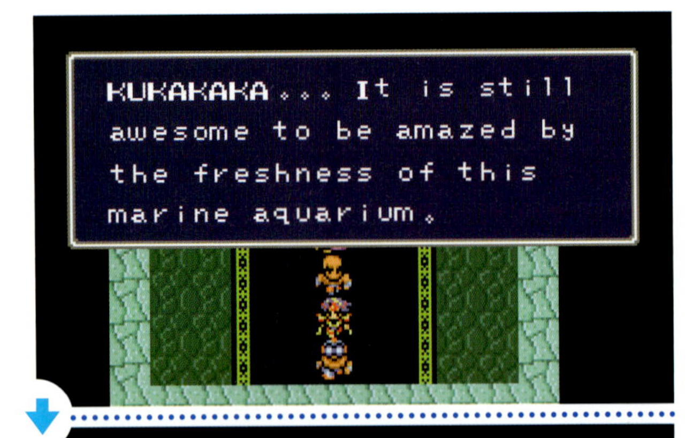

Kukakaka... Even in death I, the Water Fiend Cagnazzo, am a force to be feared! Die as you experience sheer terror! I'll be waiting for you in hell! Heheheh!

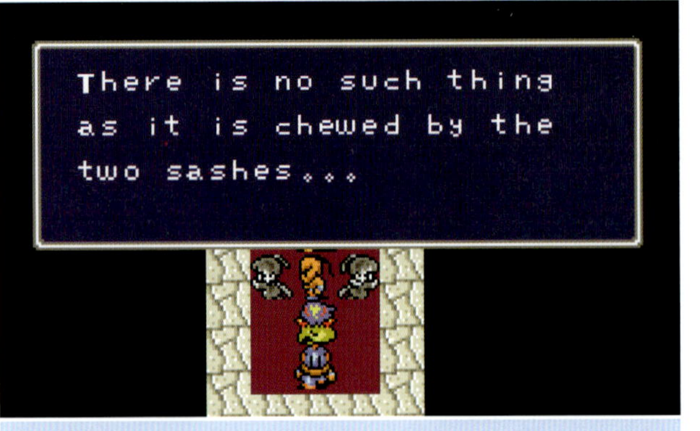

There's no effect. They've been turned to stone by their own will...

It's possible we were being controlled too.

Story 

Before the party can leave the castle, the Fiend of Water tries to kill everyone with one final trap.

Palom and Porom sacrifice their lives to save Cecil, Cid, and Tellah.

Mato Says 

I'm particularly fond of this simple line from Translation #2. It's almost as if the game has become self-aware and is using this single soldier to speak to the outside world.

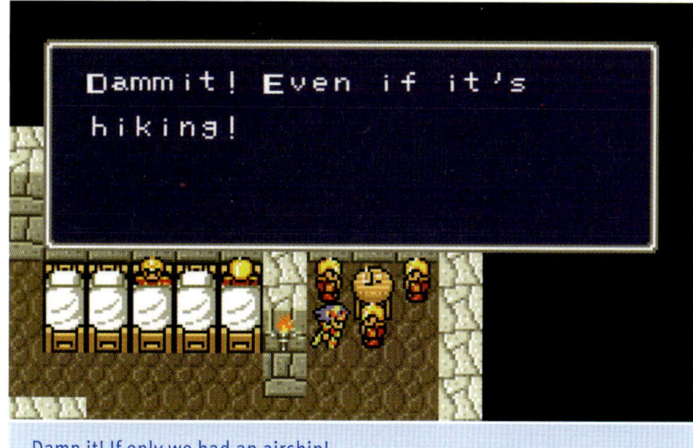

Damn it! If only we had an airship!

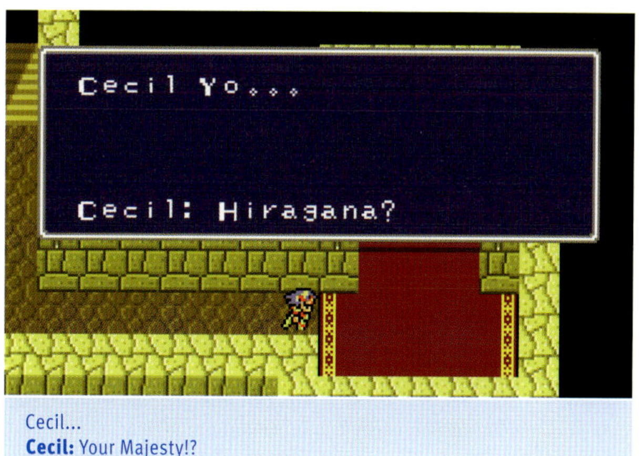

Cecil...
Cecil: Your Majesty!?

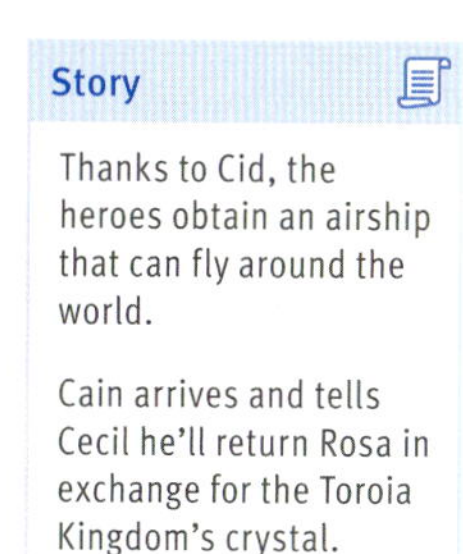
Story

Thanks to Cid, the heroes obtain an airship that can fly around the world.

Cain arrives and tells Cecil he'll return Rosa in exchange for the Toroia Kingdom's crystal.

At this point, the entire world opens up for the player to explore.

Tellah: What a dirty trick!

▲ Language Beyond the Grave

As we've already covered, *hiragana* is one of the three main writing systems in the Japanese language. There's no logical reason for *heika* ("His Majesty", "Your Majesty") to be translated as "hiragana", but the two words do share the same first letter when spelled out in English.

This seems to be a variation of the First Letter Pattern that we've seen in previous mistranslations. In this instance, the machine probably matched the first consonant of the Japanese word with the first consonant of an English word. Or maybe it followed the First Letter Pattern but changed the word *heika* to *hiragana* before translating. Whatever happened, though, it wasn't translation.

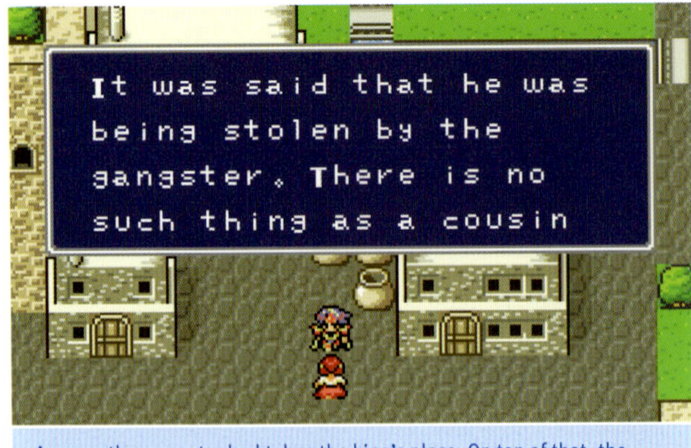

Apparently a monster had taken the king's place. On top of that, the king left no successors to follow him...

▲ Glimpses of Reason

At first glance, this machine translation looks like more random nonsense. There is some logic to it, though: a bad guy stole a person's role, and there's no family relationship left. Of course, it's easy to read it that way when you already know the original intended text – a newcomer to the game would probably be 100% baffled by this translation.

There's a well here that's been worshipped since ancient times. It's supposedly unfathomably deep.

▲ Double Trouble

Translation #2 features another peculiar pattern: phrases sometime repeat for no real reason. It's almost as if the machine gets stuck in a short loop during translation, causing individual words and whole phrases to appear twice in close succession. This mistake happens regularly enough that it's another of Translation #2's defining traits.

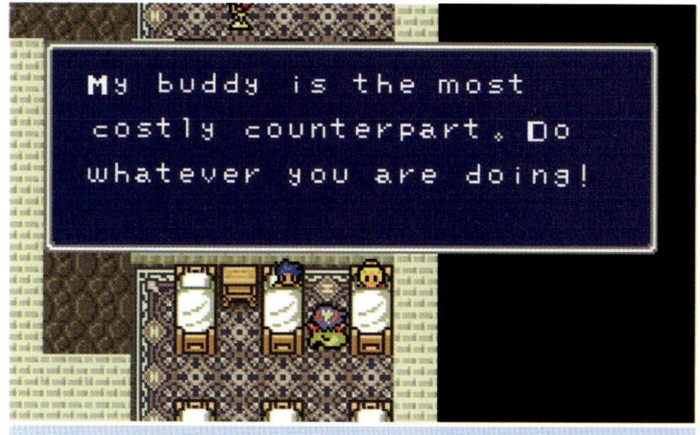

My grandma's the oldest person in the village. She knows everything!

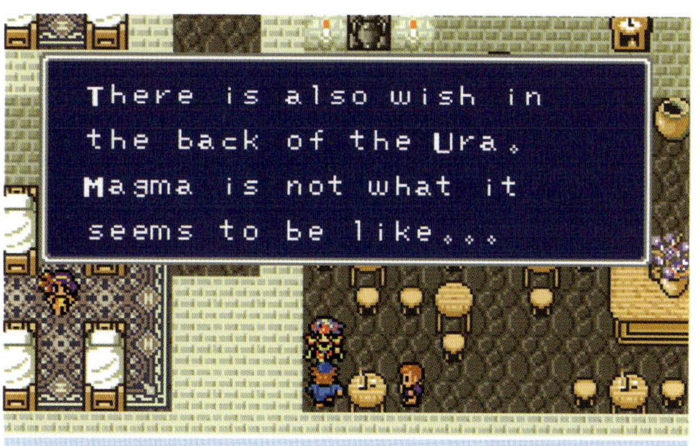

The underground has sunlight too. Sunlight known as "magma"...

Story

While visiting remote villages around the world, the heroes hear rumors about people living underground. They also learn that one of the moons has been changing color lately.

Reader Challenge

Quick! What's the difference between magma and lava? Can you name any video games that mix the two words up?

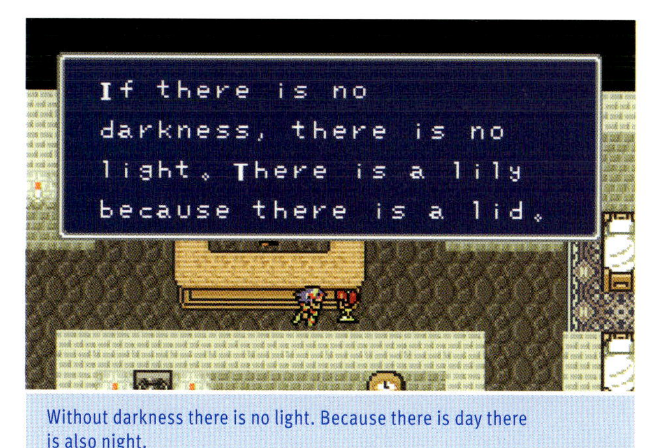

Without darkness there is no light. Because there is day there is also night.

◄ Wealth of Wisdom

A surprising number of lines in Translation #2 are worded like life advice or bizarre fortune cookie messages. Before now, I never imagined that machine translations could sound like philosophers from another dimension!

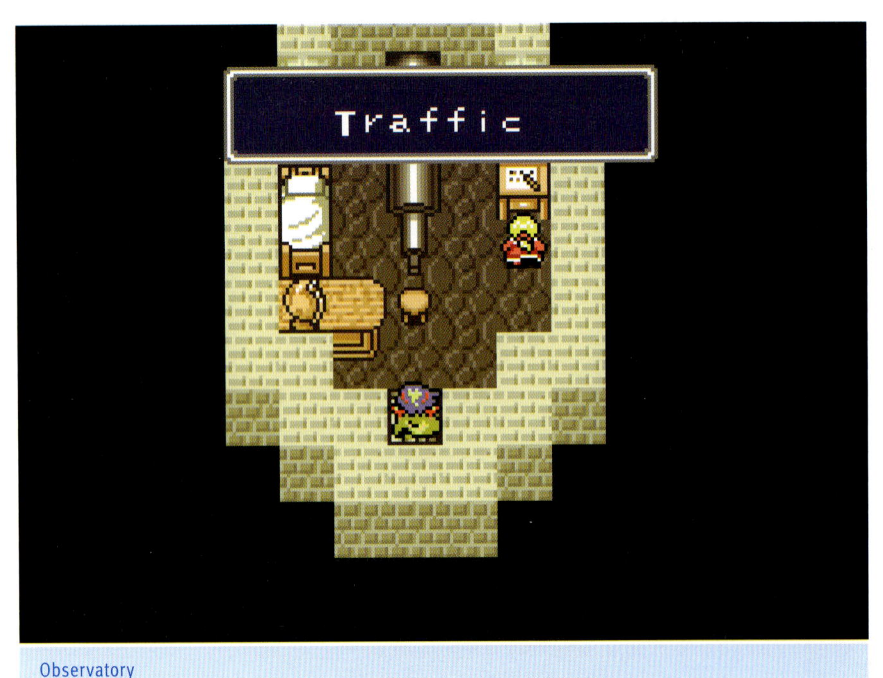

Observatory

Have you ever had a
history of mist abuse?

Is this the end of the history of Mist's summoners?

▲ MISTranslation at Work

Most of Translation #2 is so silly that it's surprising to encounter a somewhat comprehensible line. In this instance, although the translation is off, two of the key words remain intact: "history" and "mist".

I was curious to see where the rest of the translation went wrong – particularly the "abuse" part – so I did some tests with the Japanese line. I couldn't find a clear cause for any of the mistranslated words, but my tests did produce some interesting new lines, including "Whether it's time for mist abortion..." and "Have you ever been a mist boyfriend?".

Whip

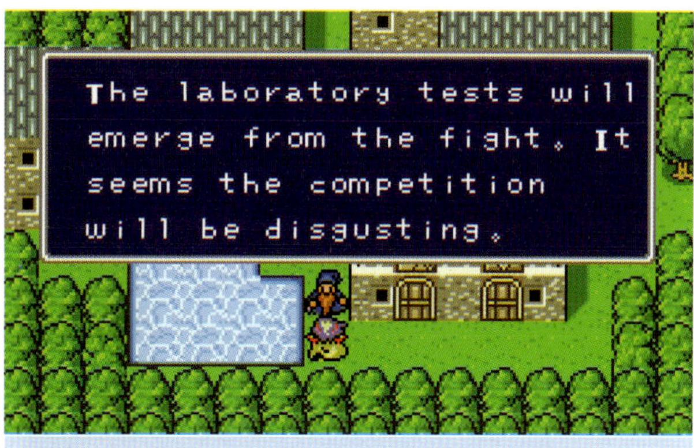

The Eidolons that we summoners call forth appear from the Eidolon Realm. It supposedly exists deep under the earth.

Mato Says

The “Eidolon Realm” goes by a different name in almost every unique translation of *Final Fantasy IV*. For simplicity’s sake, we’ll stick with this basic name.

▲ Look Again

Not all mistranslations are obvious at first glance. Some mistranslations, whether human-made or machine-made, are subtle enough to avoid immediate detection.

In this instance, a whip weapon is mistakenly called a rod, but this mistake is only clear because of the whip icon. Without it, it would’ve taken much longer to notice something was wrong.

We need the blessings of the sun to survive. They say that our ancestors were different, though.

Mithril Mountain is far to the north. With your legs, you can walk there in no time, but it's a long journey for us!

Oh, you're a human. Please try not to step on us.

▲ Piano Man

This translation features another Japanese 101-level mistake: the phrase *yama wa* ("the mountain") was misread as "Yamaha", the famous Japanese corporation that makes motorcycles, musical instruments, and more.

References to real-life companies and products are often removed during the localization process, but the machine actually *added* a reference here!

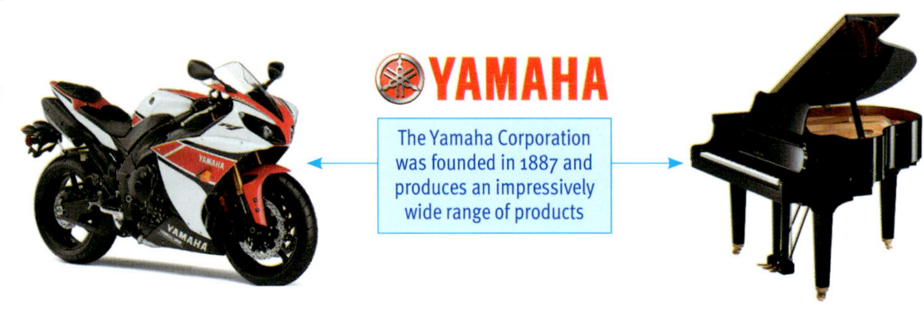

YAMAHA
The Yamaha Corporation was founded in 1887 and produces an impressively wide range of products

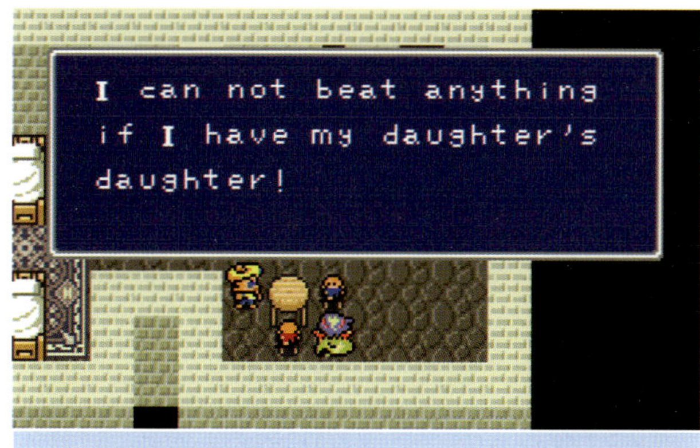

Believe me, having a pretty daughter is a source of never-ending trouble!

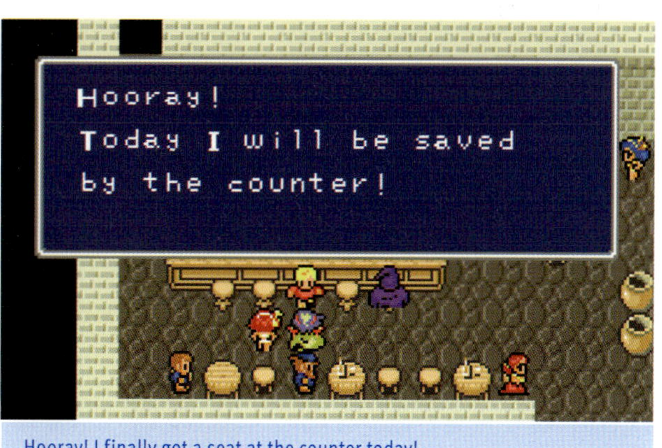

Hooray! I finally got a seat at the counter today!

Story

The heroes eventually reach the lush kingdom of Toroia, which is home to another magical crystal.

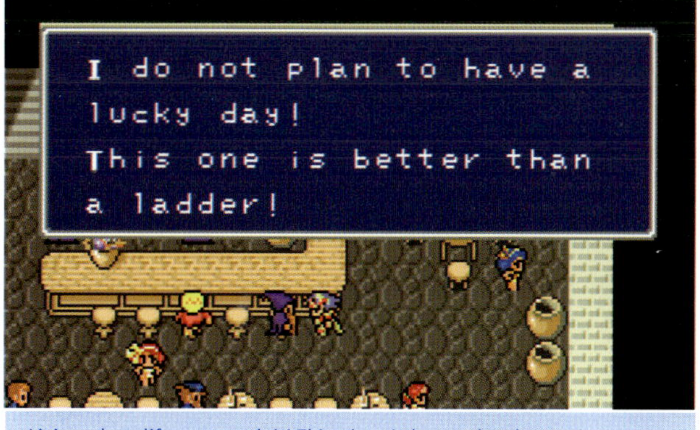

Living a long life was worth it! This place is better than heaven!

Allow me to give you folks company. What's with that look?!

Next time I want a fur.

▲ Fur Real

This simple line was almost translated correctly, but for some reason the machine replaced the final word *kegawa* ("fur", "pelt") with *kega* ("wound", "injury"). So where did the *wa* in *kegawa* go? It simply evaporated away, it seems.

Pub "King"

▲ In the Headlines

In Japanese, this secret pub is known as "Pub Ōsama". Unfortunately, the machine mistook the simple Japanese word *ōsama* ("king") for the more news-relevant name "Osama" – even though Osama bin Ladin's name is generally spelled and pronounced *Usāma bin Lādin* in Japanese. In any case, this is a good example of a ridiculous, yet plausible machine mistranslation.

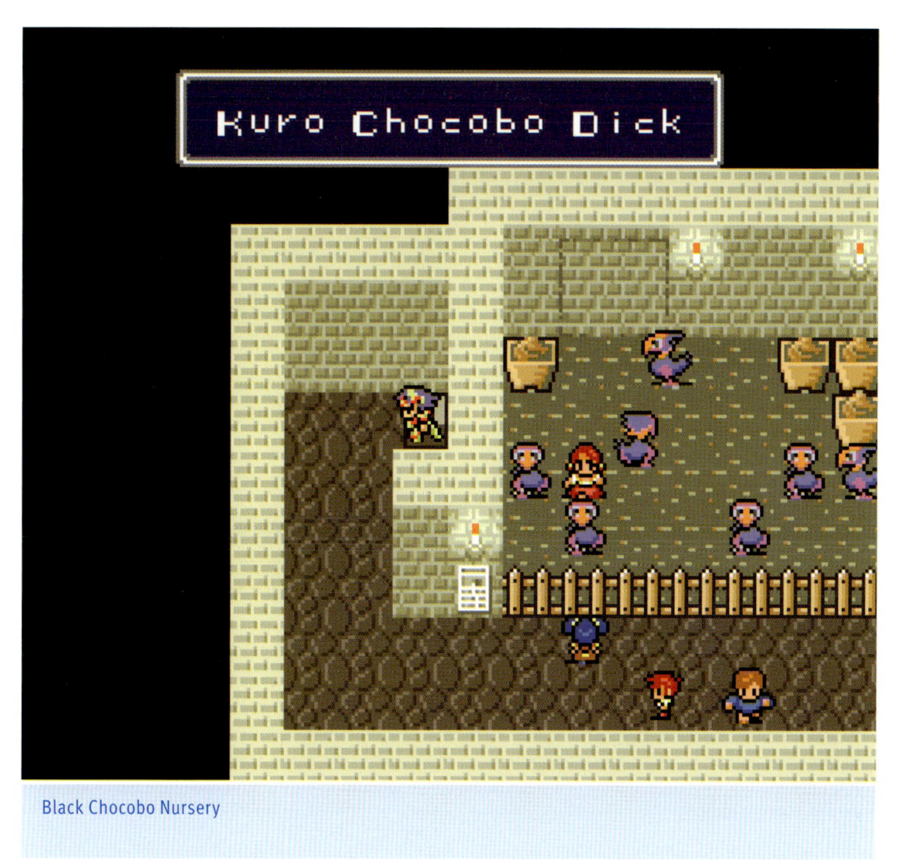

Black Chocobo Nursery

I wanna fly on a Black Chocobo too. But they get angry whenever I get close.

▲ Sound-Alikes

Although this line includes a handful of bizarre mistranslations, one of them is very simple: the Japanese word *okoru* can mean "to be angry" as well as "to occur". Context usually makes it clear which meaning is intended in everyday speech, but the machine failed to recognize the context of this line.

Kuro chocobo remembers
the harvest moat and
comes back in time. But
since the plum seems to

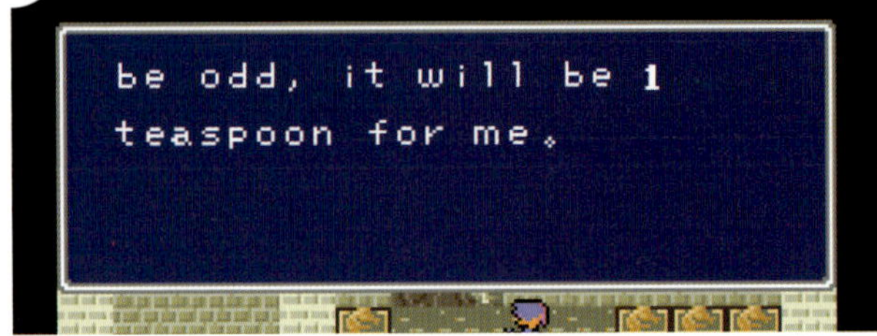

be odd, it will be 1
teaspoon for me.

Black Chocobos can remember the path they've taken and return home on their own. But they're rough-natured, so they run away after completing one round trip.

Cures petrification

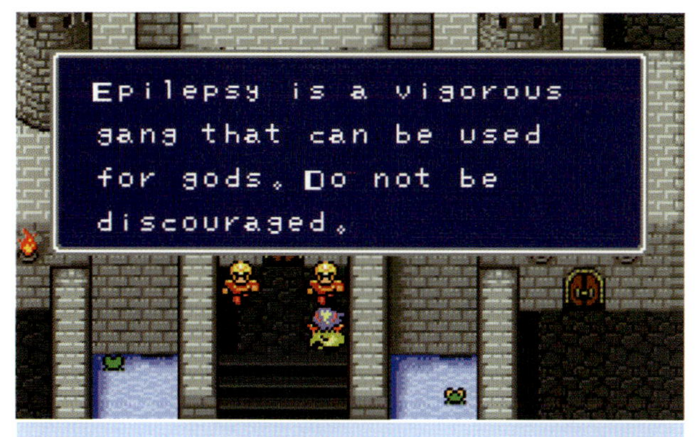

The priestesses are solemn servants of the gods. Don't do anything rude to offend them.

Do a dance? How rude! I'm a royal guard!

Story

Cecil and friends meet Prince Gilbert again in Toroia Castle, but he's too weak and injured to join the party.

The heroes ask Toroia's priestesses for their crystal. Unfortunately, it was recently stolen by the Dark Elf.

▲ Divine Inattention

In this mistranslation, the Japanese word *shinkan* ("priest", "priestess") somehow mutated into the word *tenkan* ("epilepsy") before it was translated.

This isn't an isolated mistranslation, unfortunately – these special priestesses are a regular topic of discussion during this part of the game. As a result, an entire nation in *Funky Fantasy IV* seems obsessed with epilepsy.

Me? I'm a nurse.

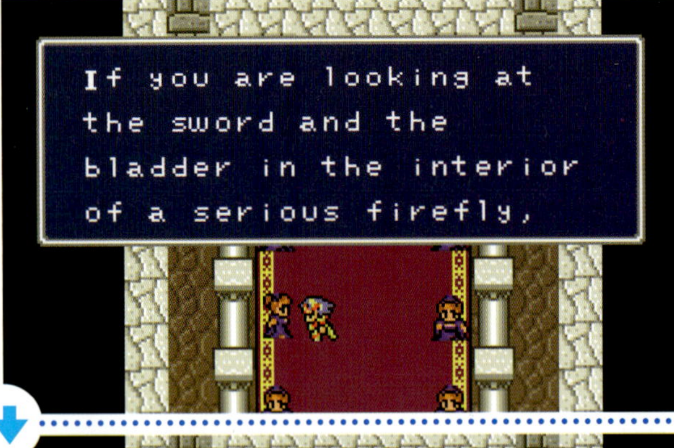

You'll be rendered immobile if you equip metallic weapons or armor inside the cave to northeast!

If you're looking for the Dark Elf, he lives in a cave on the northeast island!

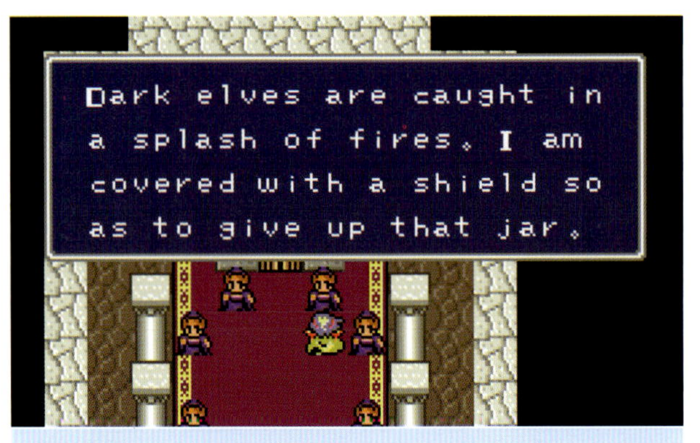

The Dark Elf is weak to metallic weapons. He has enveloped the cave in a magnetic field to compensate for that weakness.

Reader Challenge 

Quick! Use the Internet to find some fitness rooms in Thailand!

Mato Says 

These priestess characters have some of my favorite lines in Translation #2. They even outshine intentionally bad translations that are used for parody purposes!

Dark Elf: I'M IMPRESSED YOU MADE IT THIS FAR! BUT THE CRYSTAL WILL NEVER BE YOURS. DO YOU HONESTLY THINK YOU CAN DEFEAT MY MAGIC WITH THAT MEASLY EQUIPMENT?

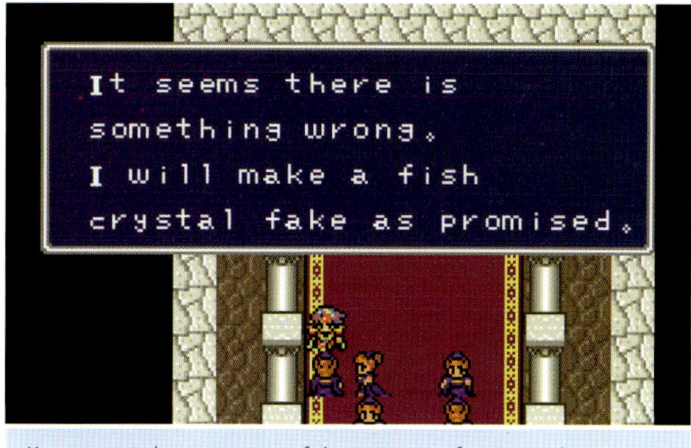

You appear to have some sort of circumstances of your own. As promised, we will lend you the Earth Crystal.

Story

The heroes spelunk the Dark Elf's magnetic cave. They defeat the Dark Elf and retrieve the crystal.

Golbez invites Cecil and companions to his tower of evil. He says he'll return Rosa if they bring the crystal to the top floor.

◂ The Trouble with Elves

In the Japanese version of *Final Fantasy IV*, the Dark Elf speaks with understandable grammar but his text is written entirely in *katakana*. This lends his speech an otherworldly vibe, similar to how robots and aliens speak entirely in capital letters in English entertainment.

In any case, this unusual text presentation completely baffled the machine translator. As a result, the Dark Elf's English translation is hardly a translation at all. Instead, the Dark Elf spews random English words surrounded by strings of untranslated Japanese gibberish.

Cures Mute

Golbez: I understand your impatience, but I want you to have your reward.

▲ Make it Rain

In this mistranslation, the Japanese word *rei* ("gratitude", "thanks") appears to have been replaced with the word "rain", simply because the two words share similar pronunciations.

Incidentally, the word *rei* has a wide variety of meanings, including "soul", "zero", "larval instar", and "example". Surprisingly, though, the machine ignored all of these definitions and chose to change the original word entirely.

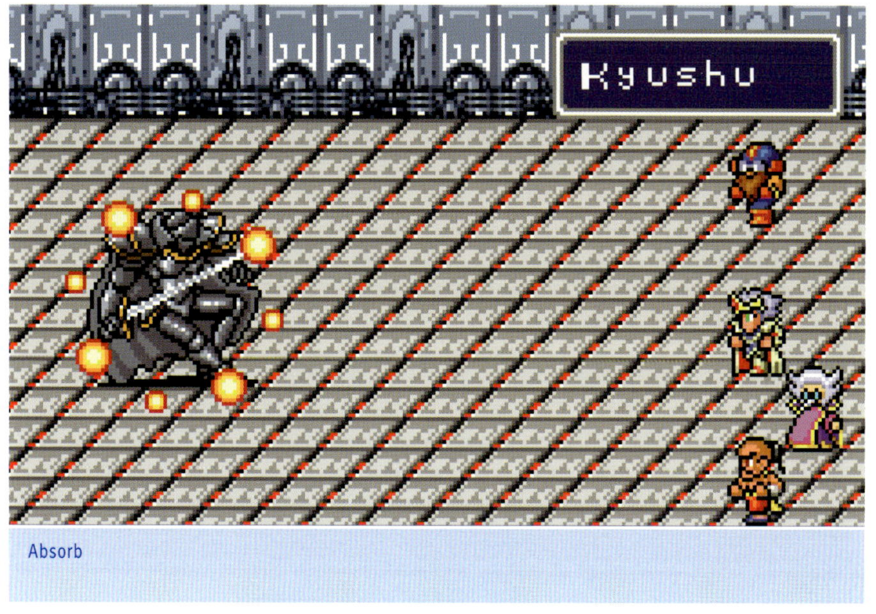

Absorb

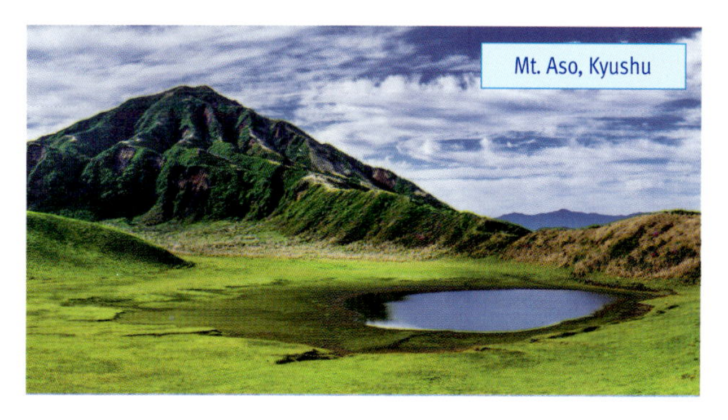

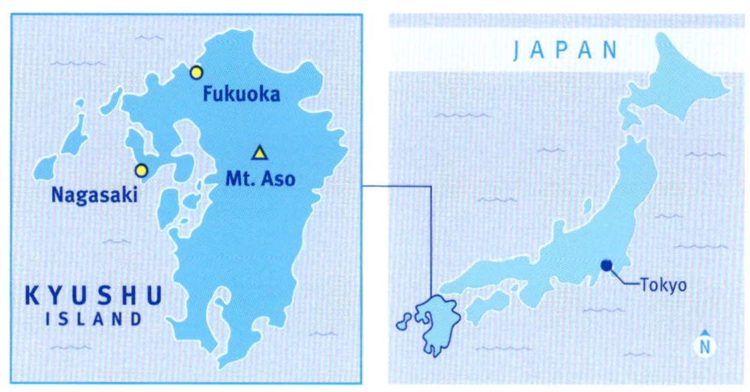

◂ Have a Nice Trip

This special technique is called *kyūshū* ("absorb") in the Japanese version of *Final Fantasy IV*. As you might suspect, the word *kyūshū* has many other meanings, including "old custom", "surprise attack", "mortal enemy", and "the 90 days of autumn". *Kyūshū* also happens to be the name of the third-largest island in Japan. Without any information about what it was translating and what the translation was for, the machine ultimately decided to go with the island name.

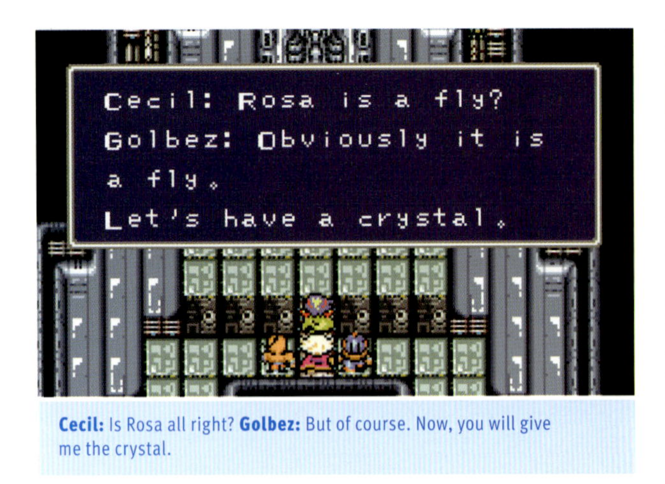

Cecil: Is Rosa all right? **Golbez:** But of course. Now, you will give me the crystal.

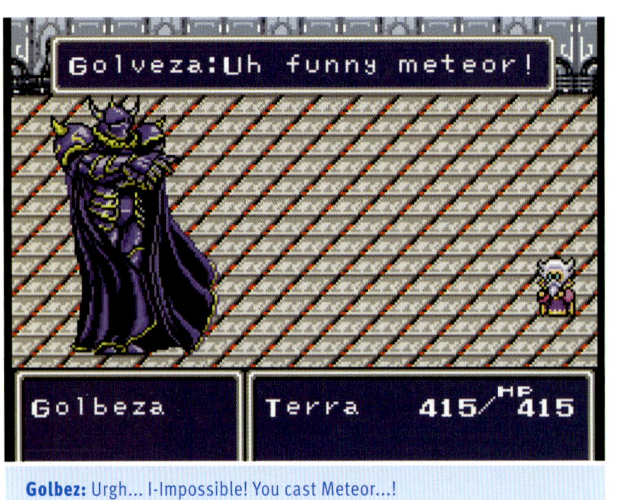

Golbez: Urgh... I-Impossible! You cast Meteor...!

Story

Golbez takes the crystal, but breaks his promise to return Rosa.

Tellah sacrifices his life to defeat Golbez and avenge his daughter, but Golbez survives and escapes with the crystal.

▲ Flying Frenzy

Characters are always getting hurt in *Final Fantasy IV*, so it's no surprise that "Are you all right?" is a common question throughout the game. For some reason, though, the machine translator replaced these lines with references to flies. The cause of this fly fascination remains a mystery to this day.

▲ Senior Discount Translation

In the English releases of *Final Fantasy IV*, the character "Golbeza" was renamed "Golbez" due to name length limitations. Professional translators often receive a list of approved name spellings before translating, but this machine translator lacked that luxury. As a result, variations of "Golbeza" appear regularly throughout Translation #2. These alternate spellings aren't necessarily mistranslations, but they do demonstrate why external terminology lists are vital to the translation process. Who knows, maybe online machine translation tools will let ordinary users supply such lists in the future.

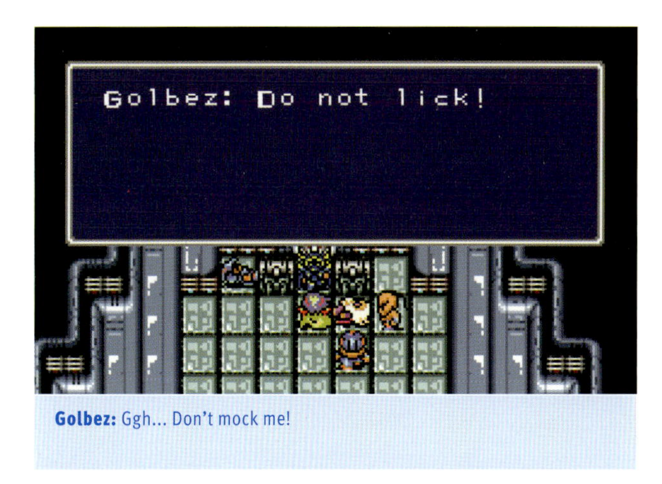

Golbez: Ggh... Don't mock me!

Golbez: Who... ARE you...? Gghaaaaaa!
Cecil: ...?

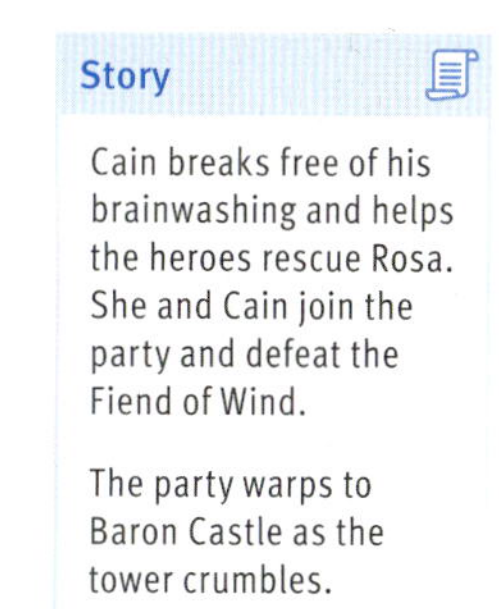

Story

Cain breaks free of his brainwashing and helps the heroes rescue Rosa. She and Cain join the party and defeat the Fiend of Wind.

The party warps to Baron Castle as the tower crumbles.

Cid: Wake up! You damn geezer!

▲ Rock the Dragon

Here, we see that Golbez's painful moan transformed into "Goku", the main hero of the *Dragon Ball* series. The cause is unknown, but appreciated.

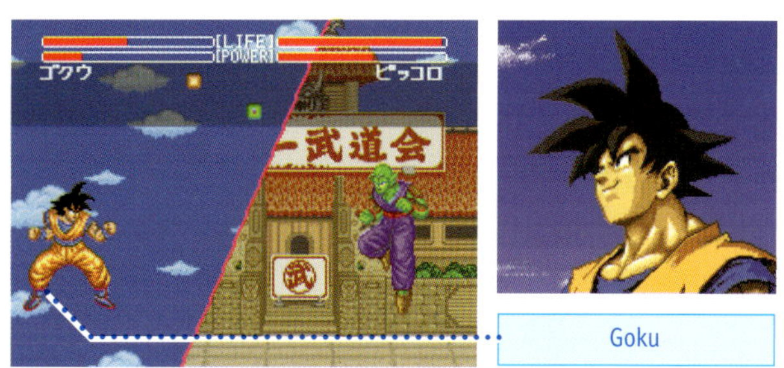

Goku

Now that the whole gang is together, I will kindly kill you all!
Cain: Heh, you aren't the only one who can fight in the air!

Cecil: I-it's collapsing! **Cain:** Damn! **Cid:** Waaah!

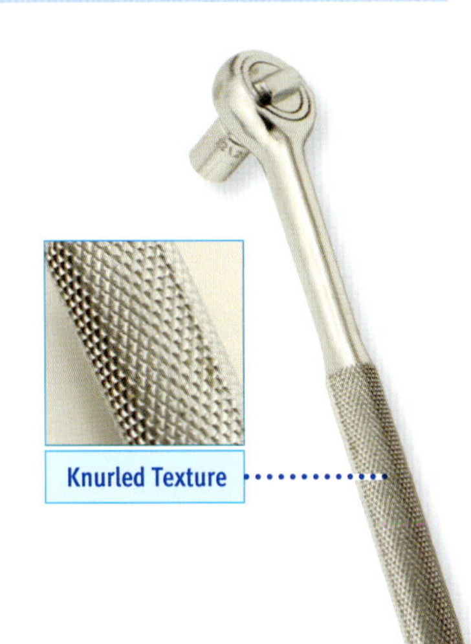

▲ Occupational Extra

There's a surprising bonus that comes with translation work: you get to learn about all kinds of stuff you'd probably never encounter in your normal, everyday life. To my surprise, it turns out that analyzing bad machine translations can be just as educational. For example, I'd never encountered the English word "knurled" before, but thanks to this weird mistranslation I've learned something new. In this way, the translation process can sometimes give back as much as you put into it.

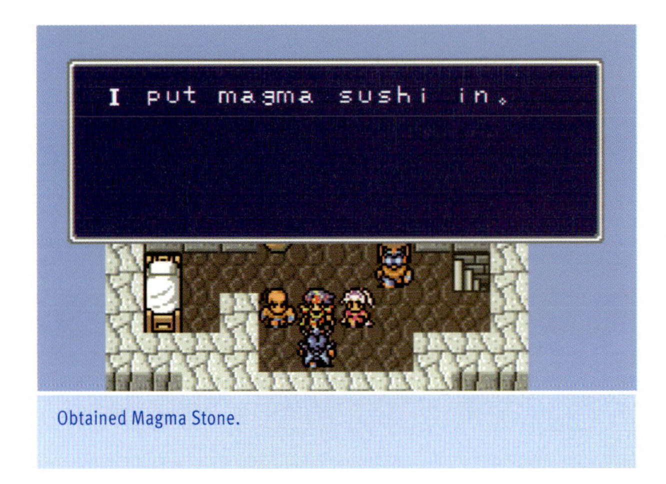

Obtained Magma Stone.

To Go Menu

Free delivery from 5:30pm (last call 9:15pm) minimum order of $30.00, within 3km

NEW!!

Flamed Salmon Sushi	**$ 4.95**
Magma Sushi	**$ 4.95**
Ebi Cheese Sushi	**$ 5.95**
Beef Tataki Sushi	**$ 5.95**
Dynamite Salad Roll	**$ 7.95**
Diced Beef Skewers (3pcs)	**$ 14.95**
Teriyaki Beef and Enoki Mushroom Roll (6pcs)	**$ 14.95**
B.B.Q. Lamb Ribs (3pcs)	**$ 14.95**
Beef Sashimi	**$ 16.95**
Fresh Baby Oyster (half dozen)	**$ 14.95**

Story

The heroes learn that Golbez doesn't have all the crystals – there are four more crystals hidden underground. The party decides to get them before Golbez does.

Reader Challenge 

If you ever find a place that sells magma sushi, try it! But only if it isn't made with fresh magma.

▲ For Sofishticated Individuals

In this simple mistranslation, the machine took the Japanese word *ishi* ("stone") and replaced it with *sushi* before translating it. Coincidentally, there's a Magma Sushi restaurant in Peru, and some sushi shops outside of Japan do have "magma sushi" on their menus!

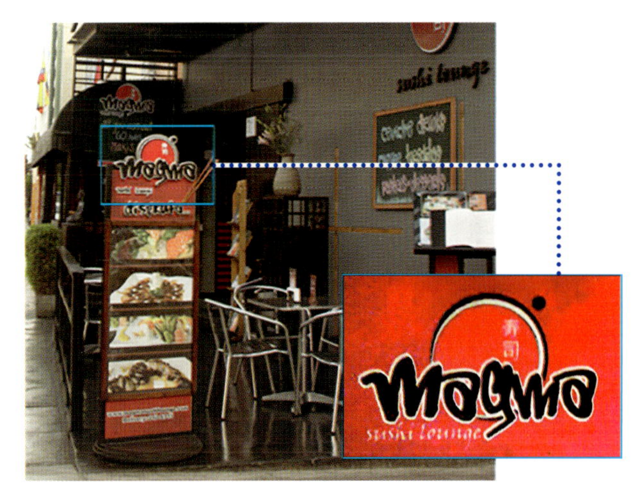

Cain: Supposedly, putting it somewhere will open the path to the Underground...

◀ Head North

This line from Translation #2 includes another reference to a real-life location. Hokkaido is Japan's large, northernmost island and is known for its harsh winters and beautiful, untouched nature.

It's unclear how Hokkaido found its way into this machine translation – there's no logical connection between it and anything in the Japanese line. I tried experimenting with the original line to pinpoint a cause, but found nothing.

By sheer coincidence, there *is* an undersea road to Hokkaido known as the Seikan Tunnel. It was even completed around the time work on *Final Fantasy IV* began!

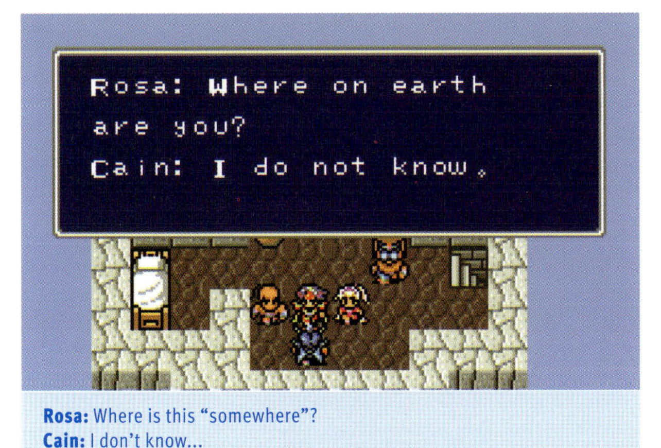

Rosa: Where is this "somewhere"?
Cain: I don't know...

But the Enterprise has gone belly up...

Story

The heroes take the airship underground. The airship crashes outside the Dwarf Castle.

Cid leaves the party to work on the airship.

Lali-ho!

◀ Larry, Ho, and Curly

The dwarves in *Final Fantasy IV* have a special greeting that has been translated into English multiple ways over the years, including "Lali-ho" and "Rally-ho". Ls and Rs are often a source of confusion when translating between Japanese and English, so it's not surprising that different people translated the greeting in different ways. The machine translator, however, chose a surprising new greeting no one ever considered: "Larry-ho". As a result of this choice, dwarves everywhere are now obsessed with some guy named Larry.

Giott: As a matter of fact, it's in a hidden chamber behind my throne!

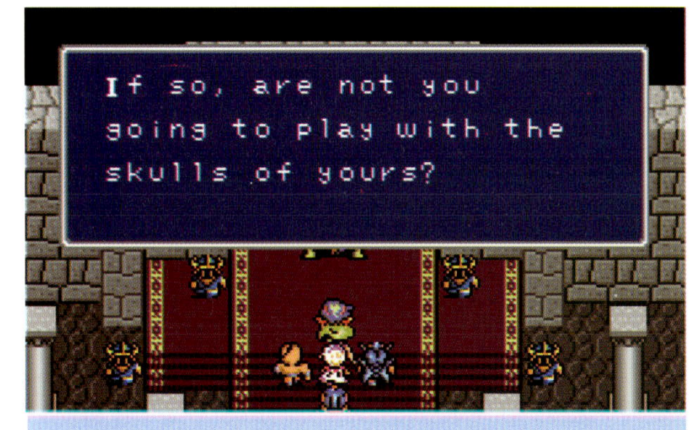

Say, perhaps you could support us with that "airship" of yours?

◀ Gangster Crib

In this mistranslation, the Japanese word *gyokuza* ("throne") was replaced with *yakuza* ("Japanese crime syndicate"), simply because of the similar pronunciation and spelling. And because *yakuza* is well known throughout the English-speaking world, the machine translator left the term untranslated. In short, the machine translator misread – or purposely changed – the source text, and then didn't even translate it afterward.

Fictional yakuza from the universally acclaimed bicycle-smashing simulator, *Yakuza 0* (PS4)

Story 

Cecil and his friends meet the Dwarf King and discuss the underground crystals.

Yang senses someone in the castle's crystal room. The heroes are ambushed when they go to investigate.

Cain: It's locked!
"Kyahohoho."

(Unshortened Text: "I'm going to have breakfast!")
The dolls merge into one!

Story

Golbez nearly wipes out the party. Rydia, now an adult, saves everyone and rejoins the group.

Despite his defeat, Golbez manages to escape with the Dwarf Castle's crystal.

The Dwarf King sends the heroes to the Tower of Babil to steal all of Golbez's crystals.

▲ Motorsports

The machine struggled to understand the whimsical Japanese laughter in this line, so it tried to break it into smaller words. We've already seen how the first part, *kya*, sounds like the Japanese pronunciation of the English word "car". The second part, "hohoho", was untranslatable, so the machine resorted to its First Letter Pattern to reach the word "hockey".

Through examples like this, it's clear that the neural network translation system indeed has a thought process of sorts, but it's still very primitive.

Come, Black Dragon!

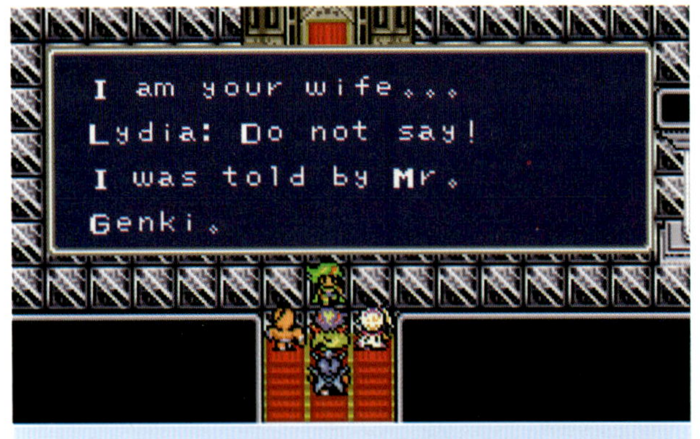

But I killed your mom... **Rydia:** Don't bring that up! The queen of the Eidolon Realm told me something.

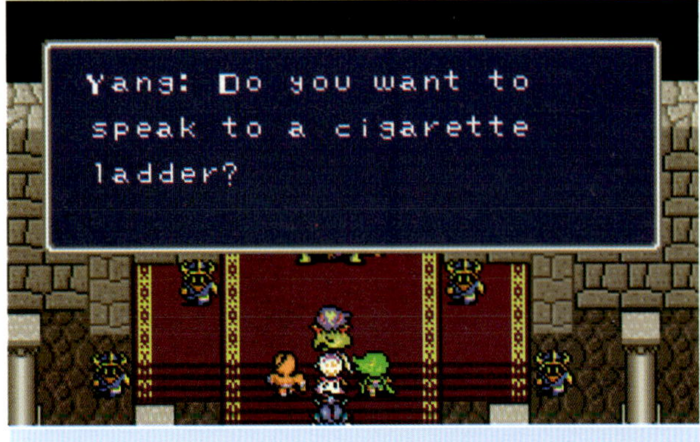

Yang: You want us to infiltrate the enemy's stronghold?!

Yang: Waiting around will get us nowhere.

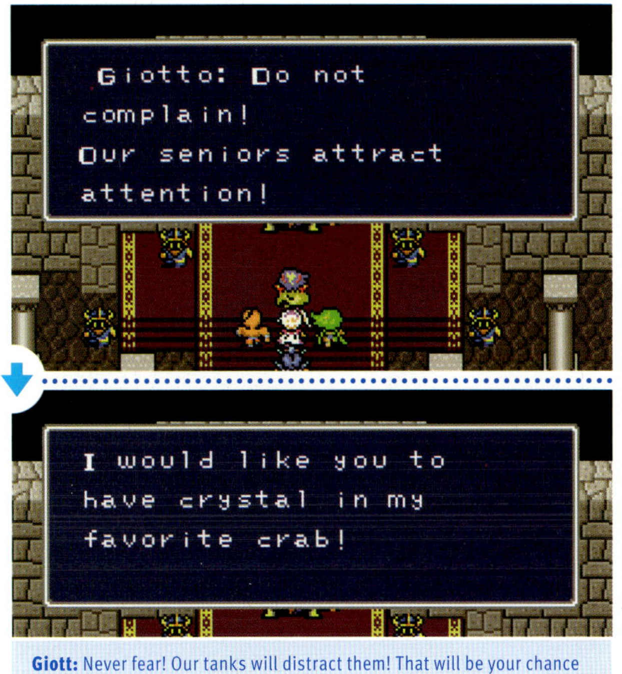

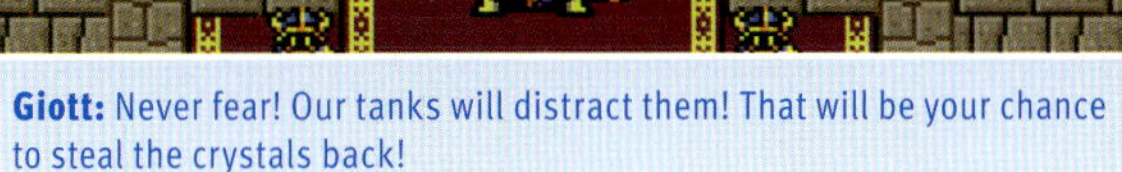

Giott: Never fear! Our tanks will distract them! That will be your chance to steal the crystals back!

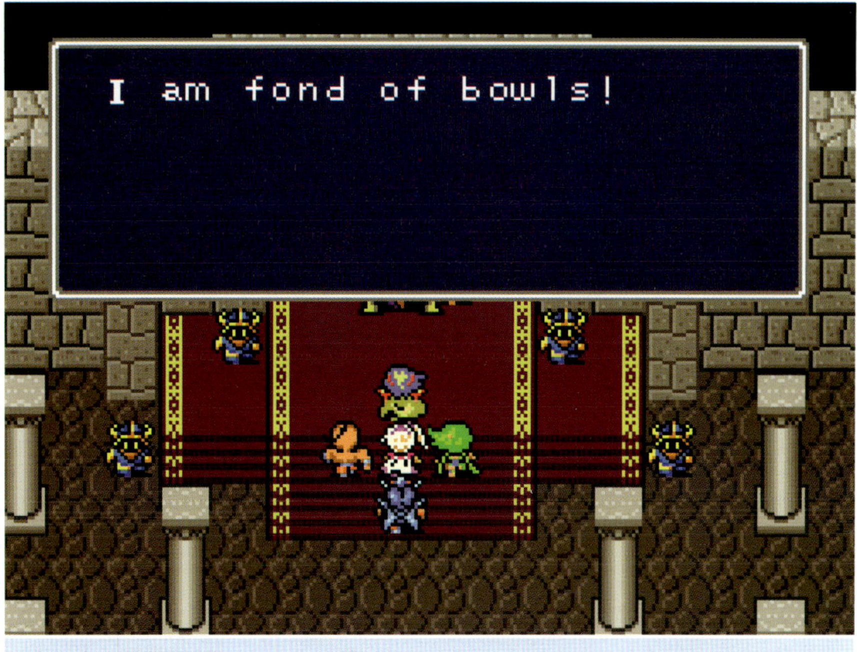

Giott: I will be praying for your success!

▲ Out of Retirement

In this line, the Japanese word *sensha* ("tank") was replaced with the word *senpai* ("a person with senior rank"), simply due to the vaguely similar spelling. The multiple meanings of the English word "seniors" obscures things further by evoking the image of senior citizens running around trying to attract attention.

Zzz...

▲ **Sleeping Like a Baby**

Onomatopoeia – the fancy name for "sound effect words" – is always tricky to translate between languages, so it comes as no surprise that machines struggle with it too.

In this example from Translation #2, the machine turned *gū gū*, the basic sound effect for sleeping or snoring, into "goo goo", which is the sound babies make in English. As a result, this exhausted, injured soldier comes across as a babbling baby.

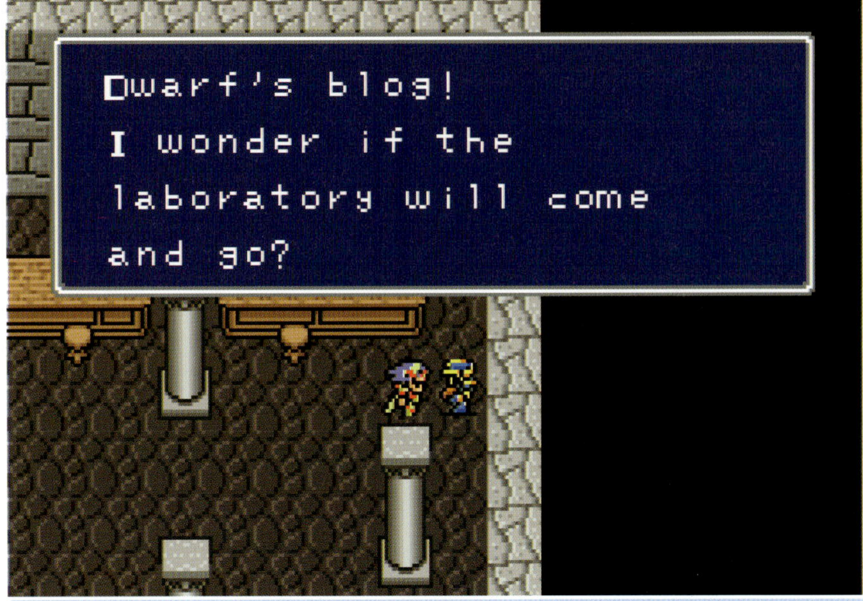

Dwarf armor is really hard! I wonder if humans can wear it properly.

▲ **Underground Web Logs**

The Japanese word *bōgu* ("protective equipment", "armor") in this dwarf's line was replaced with *burogu* ("blog"). This mistranslation is notable for two reasons: not only are the two Japanese words spelled very differently, *Final Fantasy IV* was released years before the word "blog" even existed.

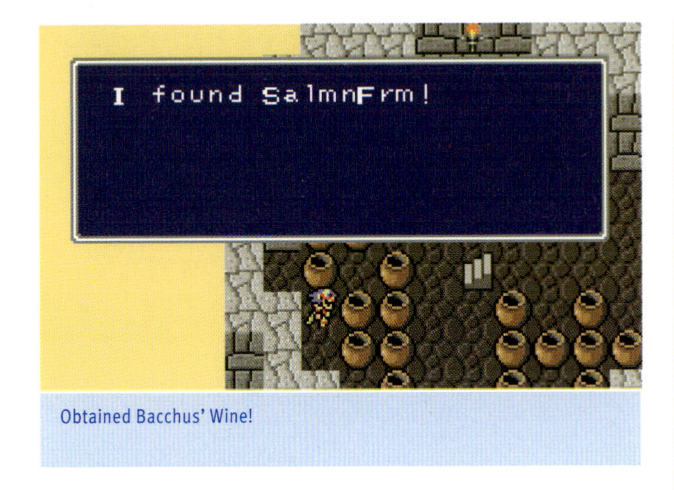

Obtained Bacchus' Wine!

▲ Fishing for Answers

The Japanese word *sake* can mean "alcohol" or "salmon". It's usually clear which meaning is intended based on context – it's very rare that you'd talk about drinking salmon, for example.

The machine translator clearly made the wrong choice in this case, though. As a result, an alcoholic drink in the game gained the "Salmon from Bacchus" name, which was then automatically compressed into "SalmnFrm". This abbreviation just made matters worse: now it looks like the item is called "Salmon Farm"!

Dwarf Castle

▲ A Bossy Fortress

This particular translation error is an understandable one: the Japanese word *shiro* ("castle") happens to sound the exact same as the command form of "to do". As a result, the full name of this area became "Dwarf Do It". Without any context or intuition to work from, the machine defaulted to assuming it was translating a full sentence.

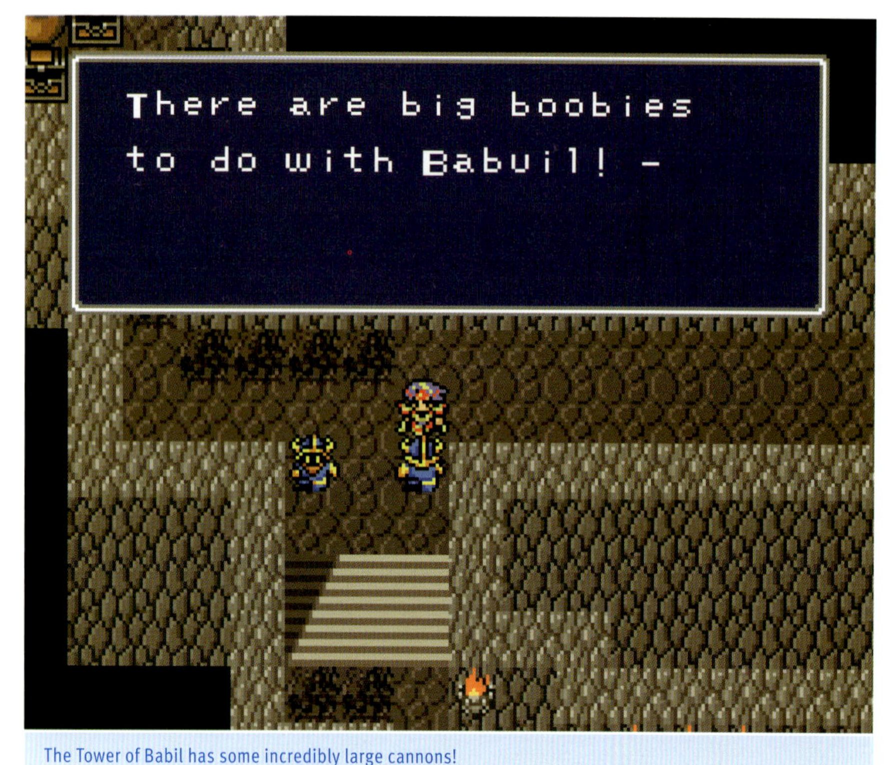

The Tower of Babil has some incredibly large cannons!

Even if Golbez isn't in Babil, Rubicante of Fire, the most powerful of the Four Heavenly Kings, is there!

▲ Tech Trouble

This unusual line isn't the result of a mistranslation but a text formatting mistake no sane human would make. Specifically, the machine turned the numeral 4 in this line into a special "control code" that caused the game to behave unpredictably. As noted before, both Translation #1 and #2 were filled with corrupted control codes like this. If I hadn't manually fixed the problems afterward, about 25% of both translations would've suffered from issues like this.

Fat Chocobo? How rude! Down here it's the God Bird!

Development Room

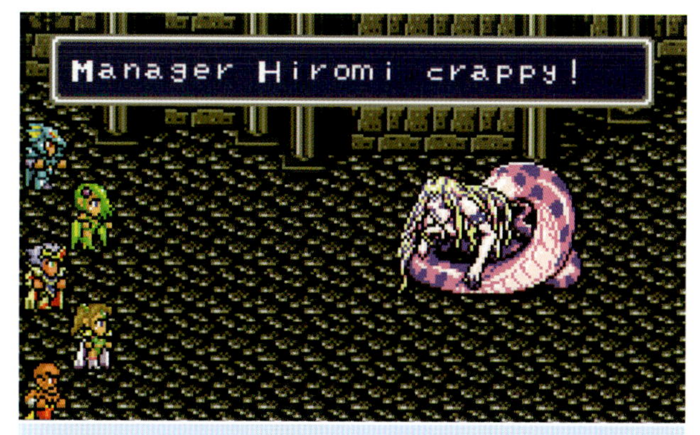

Manager: Ms. Hiromi! You're needed over here!

Please send your letters of encouragement!

Story 

A secret "Developer's Room" is hidden in the Dwarf Castle. Cecil and his companions can interact with *Final Fantasy IV*'s developers here.

Mato Says 

This secret development room was cut from the original English version of *Final Fantasy IV*, so if you're an old fan but don't recognize any of these images, check out a newer release of the game!

Found a Naughty Book!

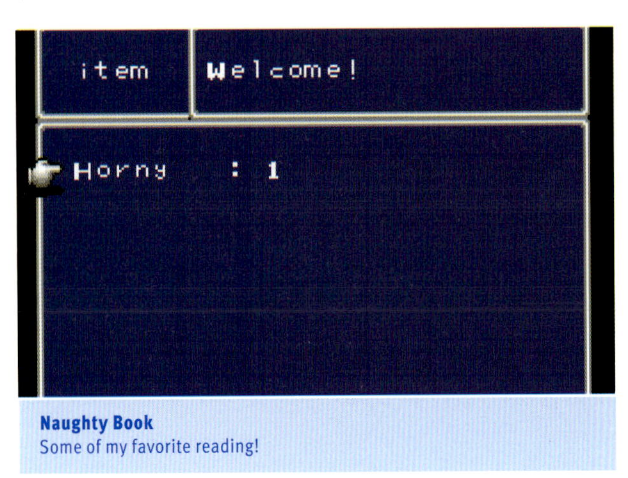

Naughty Book
Some of my favorite reading!

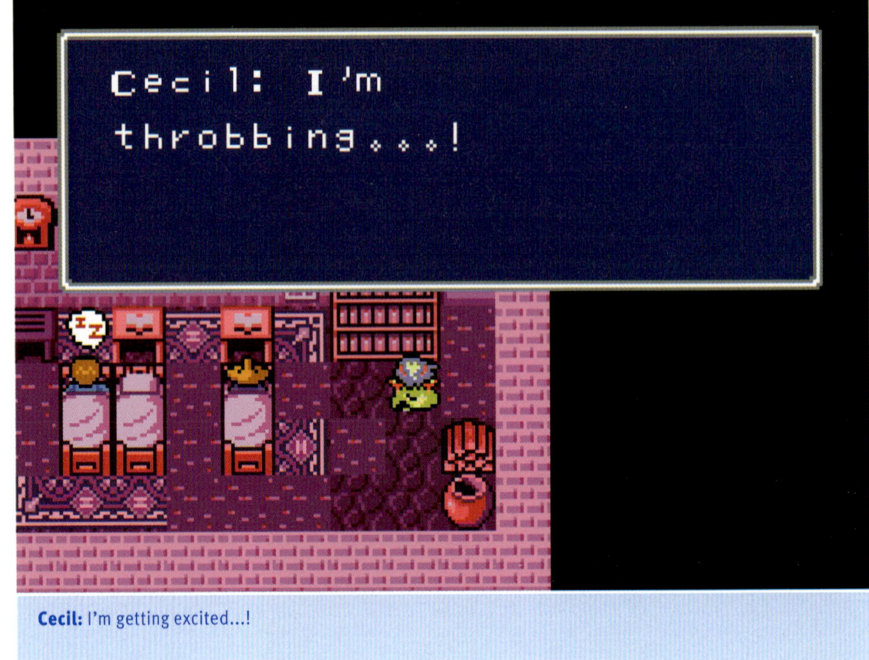

Cecil: I'm getting excited...!

▲ Doki Doki Panic

In Japanese, the onomatopoeia phrase *doki doki* refers to the sound of one's heart beating quickly, usually because something is exciting or thrilling. This heartbeat connection is what led the machine to translate *doki doki* as "throbbing" in this scene.

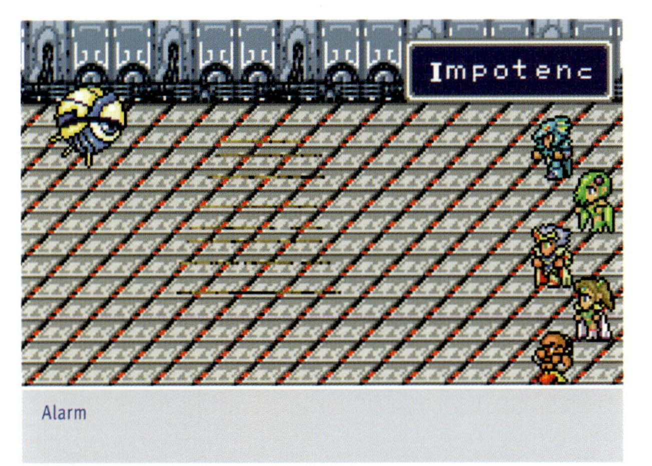

Alarm

Dr. Lugae: Urrgh! My beloved son will have your heads!

Story

The heroes sneak into the Tower of Babil to steal Golbez's crystals. They defeat a mad scientist and his killer robot.

(Unshortened Text: "Inori did not reach the train station")
The prayer did not reach the heavens

*(Unshortened Text: "**Hakase:** Yoshi show me your power!")*
Doctor: Okay! Now show 'em your power!

Reader Challenge

Quick! What was the first Nintendo game to feature Yoshi the dinosaur? Did it pre-date the original English translation of *Final Fantasy IV*?

"You cretins!" "What are you doing here!?" "Get 'em!"

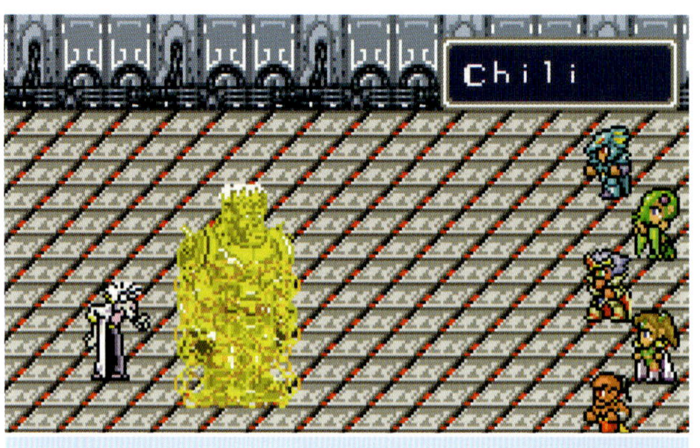

Heal

Story 

Cecil and friends try to stop the tower's cannons from bombarding the dwarves outside, but monsters block the way.

The word "chili" means different things to different cultures, so here's what American chili looks like.

▲ Outstanding in Their Field

In this mistranslation, the Japanese word *kakare* ("Charge!", "Get them!") was replaced with the word *kakashi* ("scarecrow") before it was translated into English. This is another example of the type of mistake that has come to represent Translation #2's peculiar style.

▲ Tasty Treatment

This mad doctor uses a technique called *chiryo* ("medical treatment") to cure status ailments during battle. Even though *chiryo* is a common Japanese word, the machine translator changed the word to *chiri* and then replaced the R with an L. The result: the doctor uses spicy cooking to counteract poison gas.

Cecil: It's going to explode, Yang!
Yang: ...Forgive me!

Golbez: When the cat's away, the mice will play.

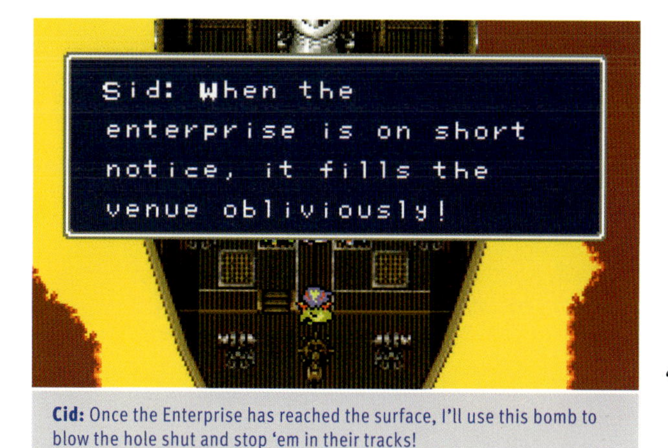

Cid: Once the Enterprise has reached the surface, I'll use this bomb to blow the hole shut and stop 'em in their tracks!

▲ The Lizard of Paws

The Japanese version of this line uses an old proverb that's literally closer to "Refreshing your life while the devil is away?". This explains why "life" is present in the machine translation, but how do "lizard" and "rice crackers" fit into the equation? The answer remains a mystery, but during my tests with the Japanese line I found alternate machine translations that are just as silly, including "Dynasty of life with dragonfly" and "Will it be the living of life during the dragon's dog?".

Story

Yang sacrifices his life to destroy the tower's cannons. The other party members are distressed.

As the heroes leave the tower, Golbez tries to kill them. Cid uses his airship to save them.

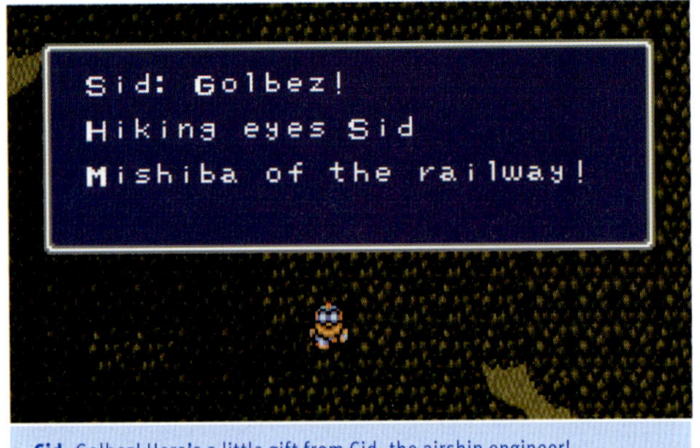

Cid: Golbez! Here's a little gift from Cid, the airship engineer!

I have become like a good otaku like a woman?
It is a father!

I've gotten as skilled as our boss, wouldn't you say? B-but don't tell him I said that!

Story

Cid sends the heroes back to the surface world, and then sacrifices his life to stop Golbez from following.

Cecil and friends head to a distant cave near the devastated kingdom of Eblana.

▲ Threat of Annihilation

In this key scene from *Final Fantasy IV*, Cid sacrifices his life for the heroes by detonating a bomb he's holding. His final words are meant to sound strong and dramatic.

In Translation #2, we see that the machine mistranslated Cid's line *and* that the machine was oblivious to the scene's importance. Skilled human translators are able to identify character motivations and the importance of individual lines, but it appears that machines still struggle with such concepts.

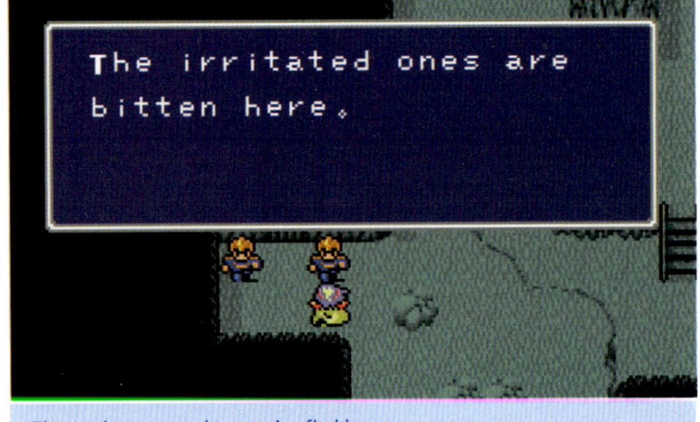

Those who managed to survive fled here.

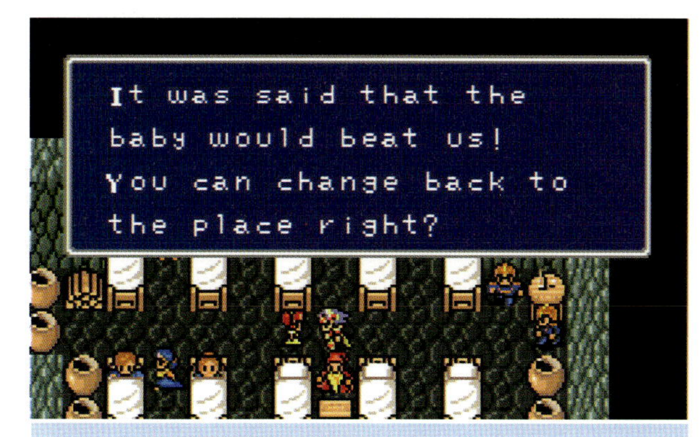

The young master said he'd beat up the bad guys for us! So we can go back to the castle soon, right?

▲ Baby Boomer

Characters refer to Prince Edge as *waka* or *waka-sama*, both of which are similar to "Young Master" in English. The machine either followed this "young" theme to arrive at the English word "baby", or it took the word *waka-sama* and replaced it with *aka-sama*, an overly respectful way of referring to another person's baby.

The Japanese word *kaeru* also appears in this line. *Kaeru* has several meanings, including "to change" and "to return to a place". Misinterpreting *kaeru* is a classic mistake that still lives on in this cutting-edge machine translation.

Transformed into a Berserk Warrior!

▲ Band of the Rock

In the Japanese version of *Final Fantasy IV*, this simple battle message appears as *kyōsenshi to kashita!* ("Transformed into a Berserk Warrior!").

Unfortunately, because the machine translator was unfamiliar with such specialized fantasy words, it misread everything in the entire line. By itself, *kyō* means "today", which explains why the translation talks about" today". It's unclear where the rest of the translation comes from, however.

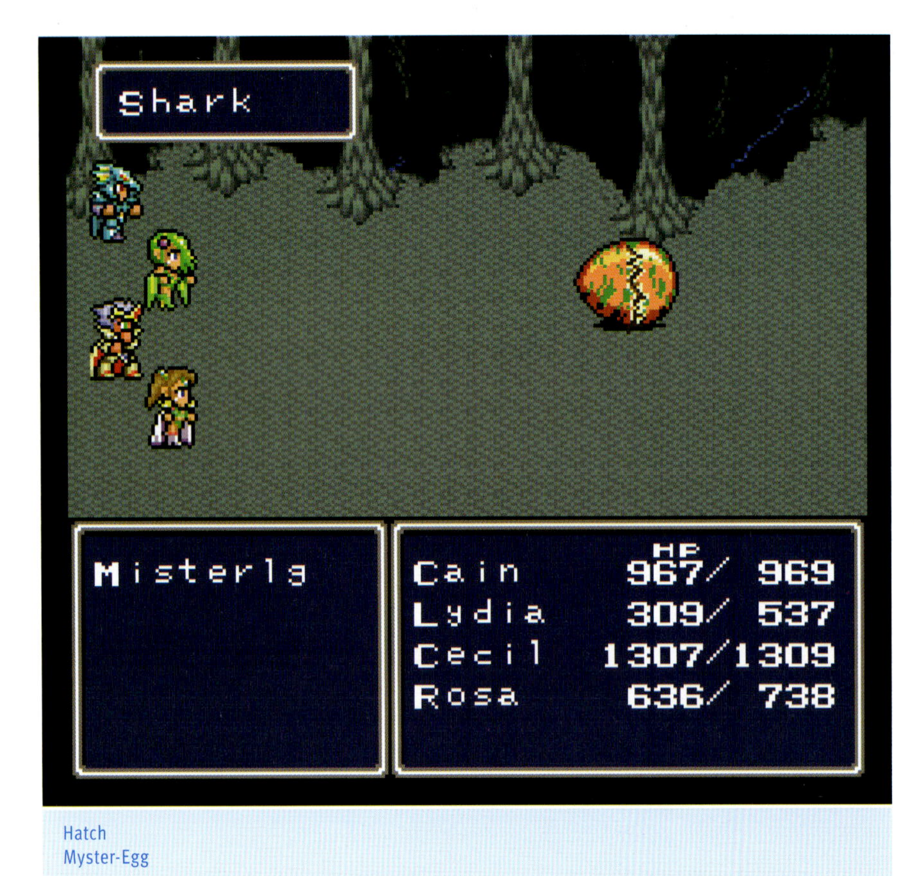

Hatch
Myster-Egg

◂ Digital Footprints

The Japanese name for this enemy, "Myster-Egg", is a clever combination of the English words "mystery" and "egg". The name sounds simplistic if you're a native English speaker, but its use of English lends the name an exotic flair for Japanese gamers.

It seems the machine was confused by this unique combination of English words, so it eventually attempted to break it into two smaller words. Unfortunately, it cut the name in the wrong place and then made the classic mistake of confusing Ls and Rs during translation. The result: a monster named "Mister Leg".

The text that appears when the egg hatches was mistranslated as well: the Japanese word *fuka* ("to hatch") was mistaken for another word with the same pronunciation that means "shark".

In all, these are mistakes that a moderately experienced human translator would have no trouble with. Recognizing others' creativity is a skill that comes naturally for us, but not necessarily for machines.

Flame Torrent

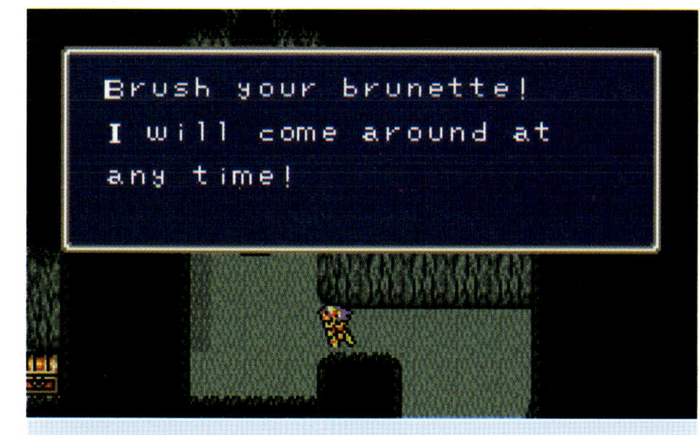

Go polish your skills! I will face you in battle anytime!

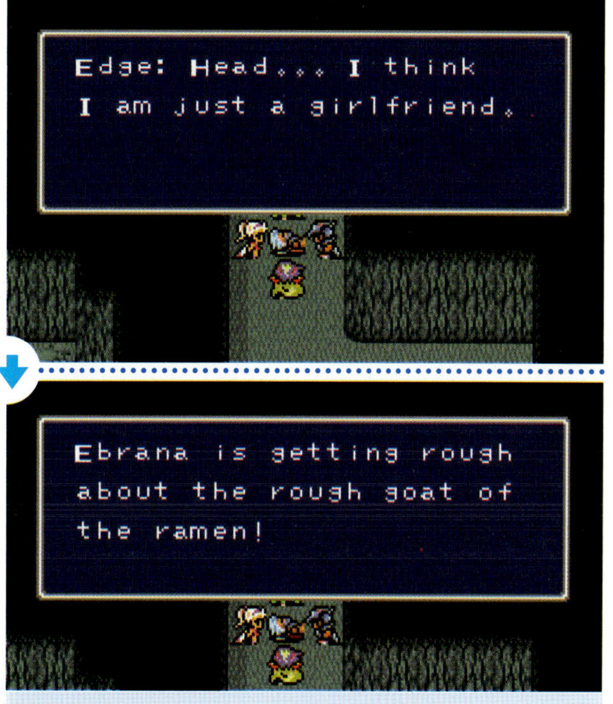

Edge: Heh... Don't mistake me for a spoiled little prince. I've inherited the Eblana royal family's ninja techniques!

Story 

The heroes head deep inside the cave to find a way back into the Underground.

Along the way, they meet Edge, the prince of Eblana. He joins the party after he loses to the Fiend of Fire.

Mato Says 

From time to time, ask yourself this question: "If I wasn't familiar with *Final Fantasy IV* or Japanese, would this translation be any more helpful than the original game with Japanese gibberish squiggles?"

In fact, ask yourself that whenever you see any bad translation!

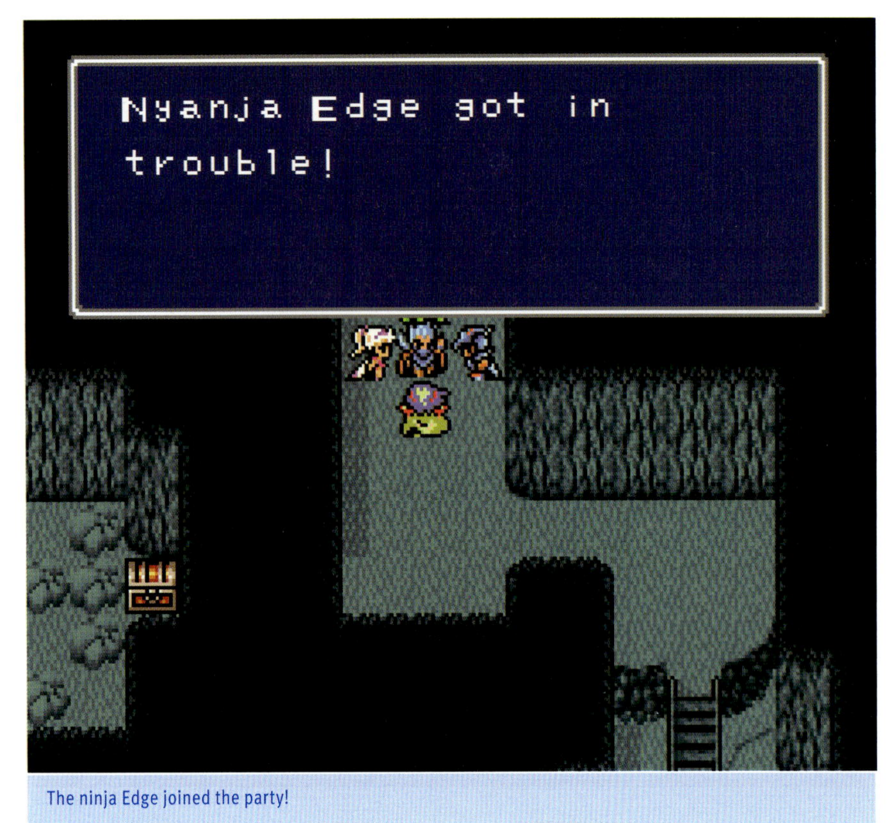

The ninja Edge joined the party!

◀ You Gotta Be Kitten Me

The word *ninja* in this line somehow morphed into *nyanja* during translation. *Nyanja* isn't an actual Japanese word, but *nya* is the sound that cats make in Japanese, so *nyanja* evokes the image of a cat ninja that you might find in a children's TV show. In fact, it turns out there *is* an obscure cat ninja named "Nyanja" in the famous *Anpanman* series!

Unfortunately, there's no way to ask the machine translation system why it chose *nyanja* here, but my gut instinct says the reason was unrelated to the *Anpanman* character.

Edge
Ninja

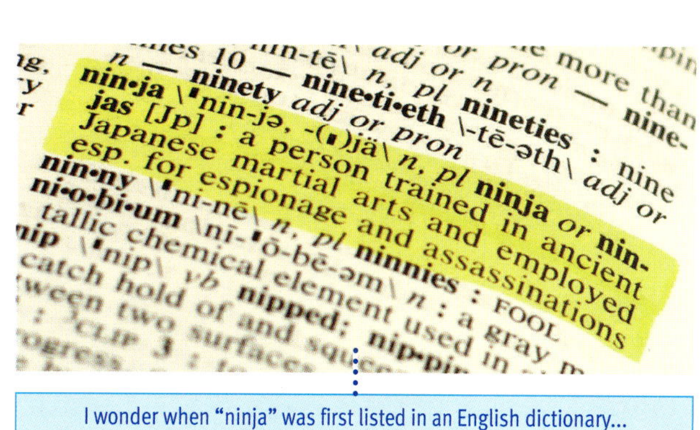

I wonder when "ninja" was first listed in an English dictionary...

◀ Bootleg Ninja Fighter

Edge is classified as a *ninja* on the Japanese menu screen, but as "A Boy" in Translation #2. It's surprising that the Japanese word *ninja* would cause so much trouble for a state-of-the-art machine translation system – the word is well-known outside of Japan and even appears in all of the big-name English dictionaries.

So, why is *ninja* so difficult to translate, and how did the machine turn it into "A Boy"? I spent an afternoon digging deeper into the problem, but it's still a mystery beyond my understanding.

Story 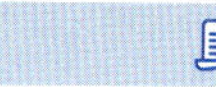

Edge uses his ninja skills to get the heroes back inside the Tower of Babil. The party explores areas of the tower that were previously inaccessible.

Reader Challenge

Quick! Think of five more Japanese words that are now part of the English language!

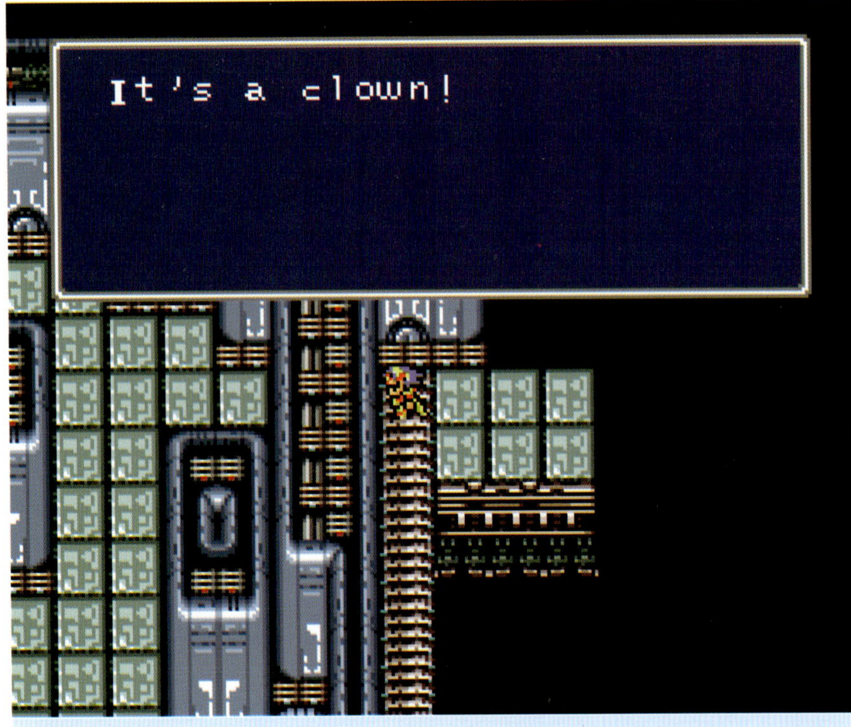

Time for my wall-piercing technique!

Edge: Wait! Mom!

▲ Give a Hoot, Don't Transmute

In an emotional scene, Edge yells out to his disfigured, dying parents. In this particular line, he calls out to his mother with the word *ofukuro* ("Mom"), which the machine misinterpreted as *fukurō* ("owl"). Although it's clearly a translation error, it's not as illogical as other mistranslations in Translation #2. In fact, I'd bet money that amateur human translators make the same mistake from time to time too.

Edge: All right! Now let's go thrash Golbez!

I leave the young master in your hands! Everyone! Let us leave!

▲ Do or Do Not

The Japanese version of this line begins with Edge shouting the word *yossha* ("all right!", "way to go!"). Although it's a common, everyday phrase, the machine unexpectedly morphed it into *yoda*, which isn't even a Japanese word. As a result, instead of showing excitement, Edge appears to be yelling about a green *Star Wars* character.

▲ Leap Year

As we've seen before, the Japanese word *kaeru* has several unrelated definitions. Besides "to change" and "to return to a place", *kaeru* has many more meanings, including "to turn around", "to exchange", "purchasable", "to replace", "to hatch", and "frog". With so many different meanings, it's no wonder that the machine translator made the wrong choice here.

Story 

The heroes eventually find Edge's parents, but they've been transformed into horrifying creatures.

Edge goes into a rage and defeats the Fiend of Fire with the party's help.

Mato Says 

Incidentally, these multiple meanings of the word *kaeru* are potentially why the transformed frog character in *Chrono Trigger* is known as "Kaeru" in Japanese.

Cecil: A pitfall?!

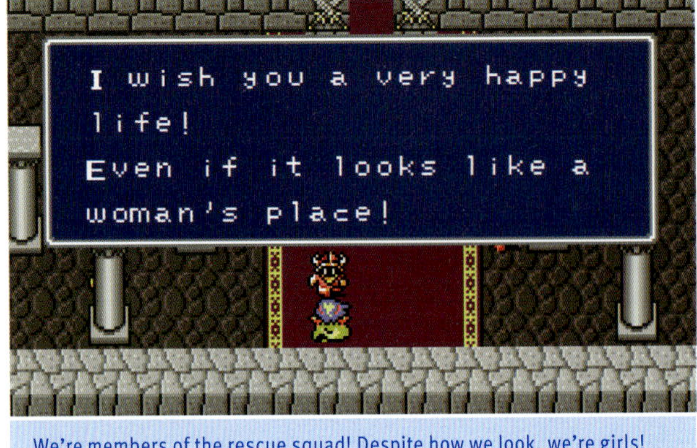

We're members of the rescue squad! Despite how we look, we're girls!

Cecil: Is this the enemy's new airship model?
Edge: Let's use it to break outta here!

◂ Mistranslation Express

In this mistranslated line, the Japanese word *shingata* ("new model") was replaced with *shinkansen* ("bullet train"). The word *shinkansen* was left untranslated, though, because it's slowly becoming a part of the English language in some parts of the world.

The word "shrine" appears in this machine translation as well, but its origin is less clear. After experimenting with the Japanese phrase a bit, my best guess is that the machine took the "sh" from the start of *shingata* or *shinkansen* to arrive at "shrine". Some alternate machine translations for the first sentence included "Shigeru air sky boat" and "shinkansen shinkansen".

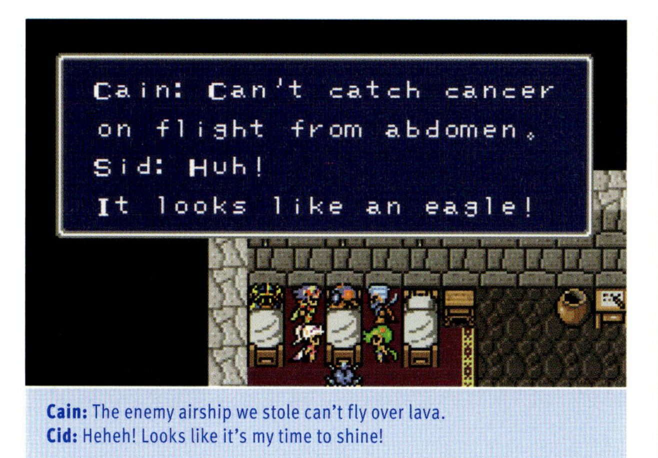

Cain: The enemy airship we stole can't fly over lava.
Cid: Heheh! Looks like it's my time to shine!

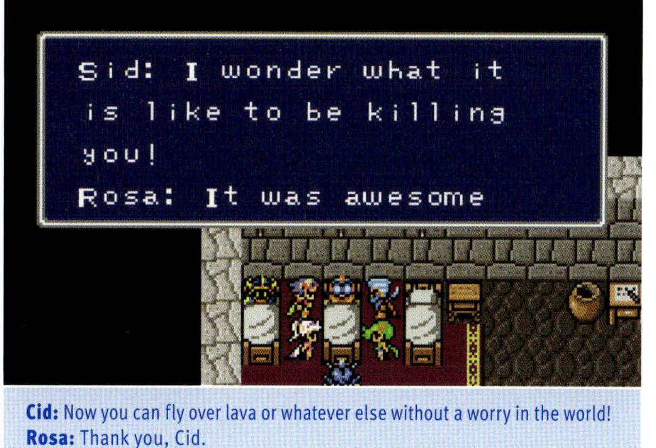

Cid: Now you can fly over lava or whatever else without a worry in the world!
Rosa: Thank you, Cid.

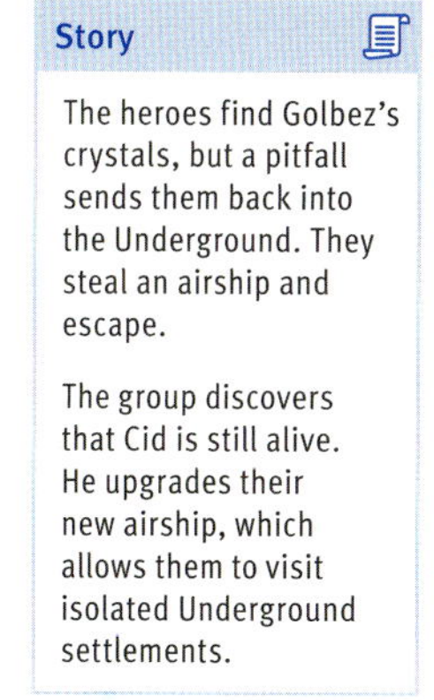

Story

The heroes find Golbez's crystals, but a pitfall sends them back into the Underground. They steal an airship and escape.

The group discovers that Cid is still alive. He upgrades their new airship, which allows them to visit isolated Underground settlements.

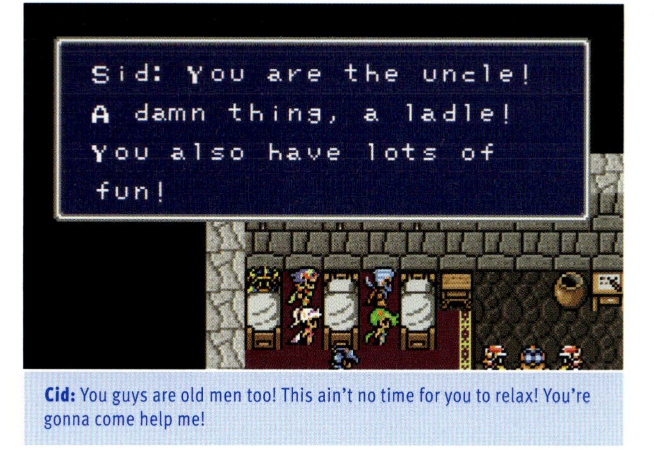

Cid: You guys are old men too! This ain't no time for you to relax! You're gonna come help me!

◀ Family Ties

In Japanese, it's common to refer to strangers using family terms. For example, if you're a little kid and you're talking to a teenage boy, you might call him "Big Brother". If you're talking about a middle aged lady you see across the street, you might refer to her as "that aunt". Similarly, middle-aged to older men are often called "Uncle".

Unfortunately, the nuances of these relationship terms are difficult for machines to grasp, particularly when there's no context to reference. Here, we see that the machine literally translated the word for "uncle" instead of the intended "older man" meaning.

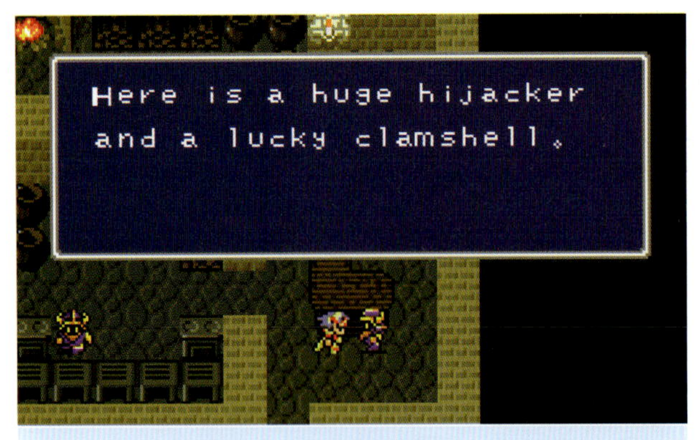

This is the home of Kukuro, the world's greatest blacksmith, lali.

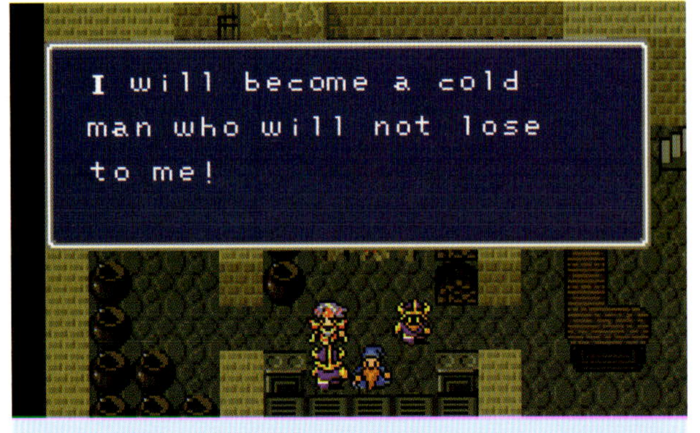

I'm gonna be a blacksmith just as good as the master!

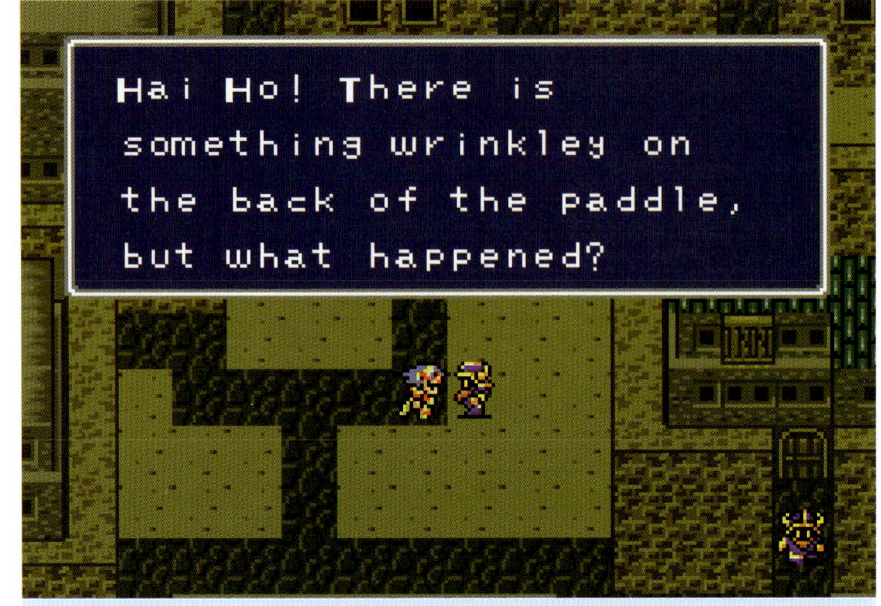

Hi-ho! There's been a big ruckus over at the castle lately. Did something happen?

▲ All Y'all's Accents 'n Dialects

The citizens of this remote village speak with a heavy Japanese accent commonly associated with rural communities. Unfortunately, their accent is so heavy that the machine translator couldn't understand a single word they said.

They say the oceans on the surface are blue. That's creepy!

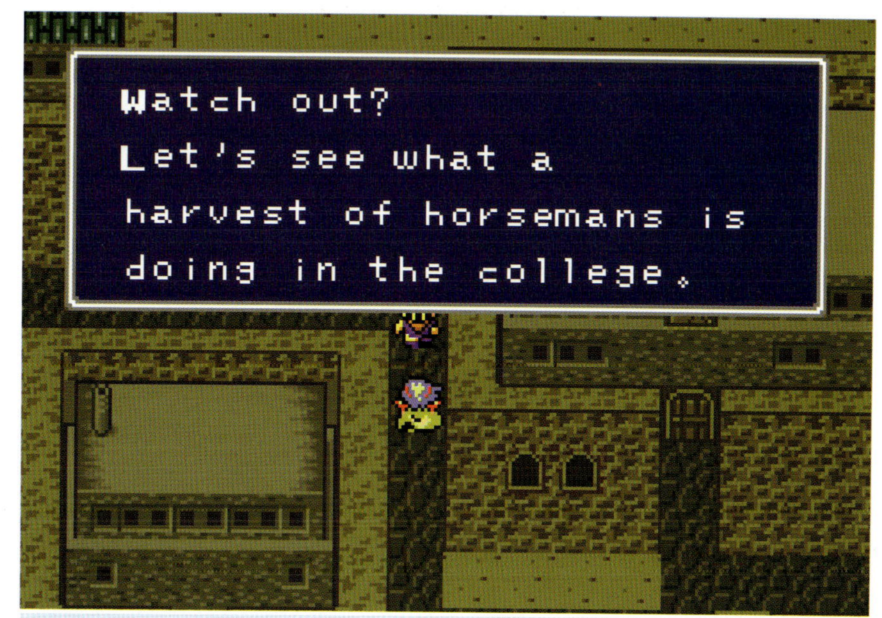

Did ya know? On an island to the northwest there's a cave that apparently leads to the Eidolon Realm.

In the far northwest corner there's a cave where fairies called Sylphs live!

▲ Linguistic Collapse

The machine translator obliterated almost every line from this part of *Final Fantasy IV*. It's hardly surprising, though, as the combination of this town's rural accent, dwarf dialect, and fantasy words would be a challenge for many human translators as well.

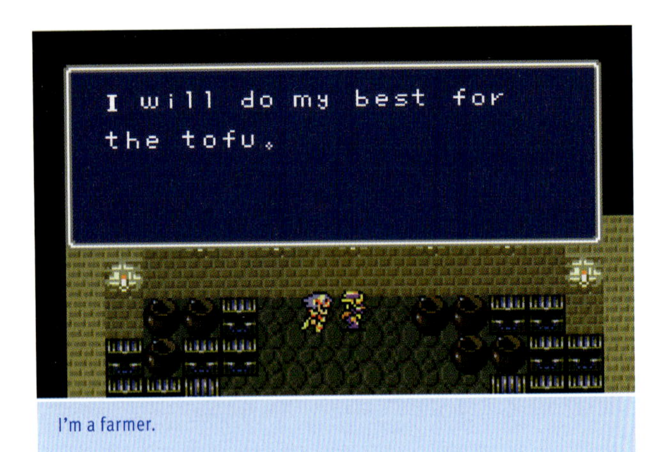

I'm a farmer.

Rydia: I passed through here when I left the Eidolon Realm!

Story

Now that the Underground is fully accessible, several side quests become available.

One major side quest takes the heroes to the Eidolon Realm, where all of the summoned monsters live.

▲ That Strange Bean Curd

Even this super-simple Japanese sentence was too much for the machine translator to handle. Here, we see that the Japanese word *nōfu* ("farmer") was replaced with the word *tōfu* ("tofu") before it was translated. Why it changed the word at all remains a mystery.

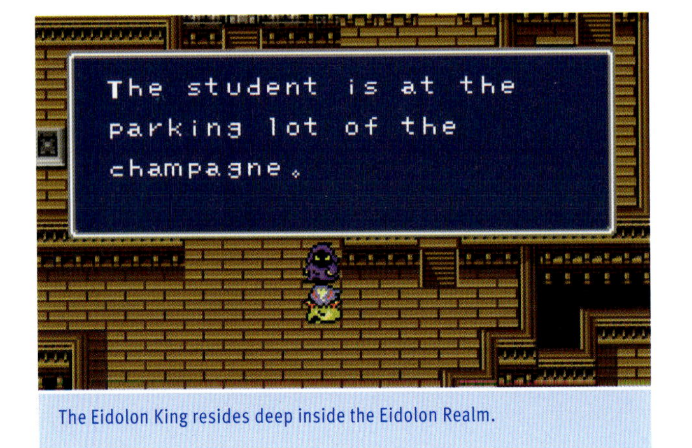

The Eidolon King resides deep inside the Eidolon Realm.

This is the Eidolon Library. It's a treasure trove of knowledge!

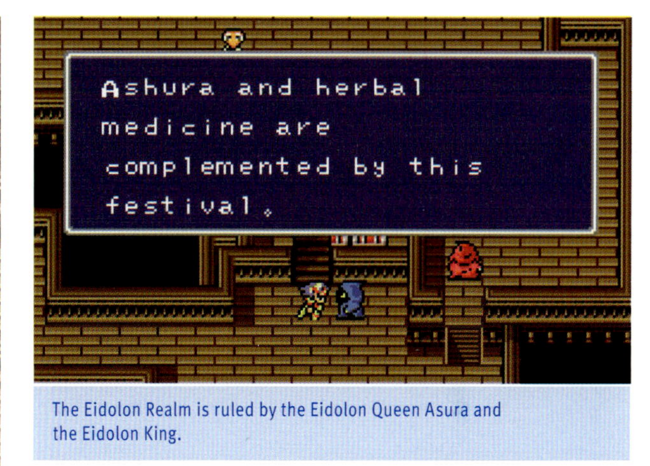

The Eidolon Realm is ruled by the Eidolon Queen Asura and the Eidolon King.

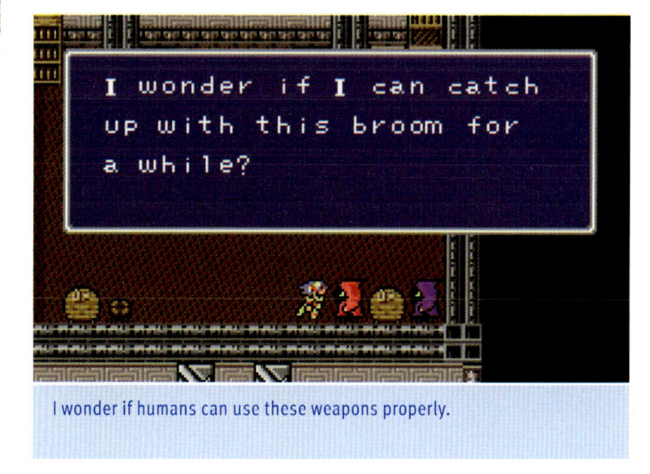

I wonder if humans can use these weapons properly.

▲ He Who Smelt It

In this mistranslation, the Japanese word *chishiki* ("knowledge") was distorted into the near-nonsensical *chisaki* and then left as-is.

After a few tests with the Japanese line, I also found that the machine consistently translates *hōkō sa* ("a treasure trove") as "smelt". I don't know what causes this consistent behavior, but I believe the machine morphs the Japanese word into something else, translates it into an English word that starts with "s", and then chooses a new word that also starts with "s".

Kotetsu

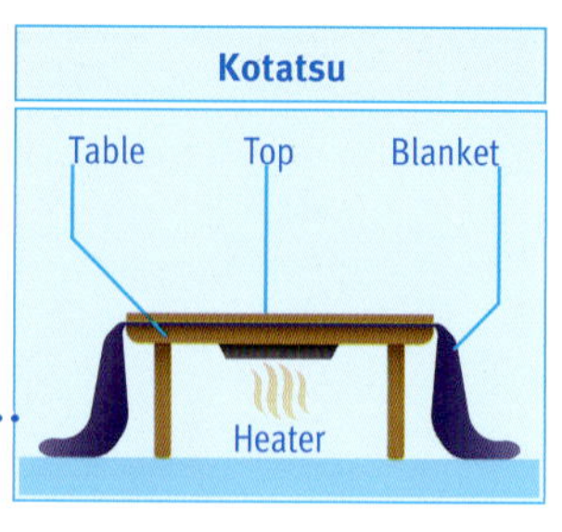

◀ Weapons of Mass Relaxation

Like many ninja swords in *Final Fantasy IV*, the Kotetsu is named after a real-life, renowned Japanese swordsmith who lived centuries ago. It seems the machine translator was unaware of this connection, but to be fair, background information like this is often missed by human translators too. In this case, though, the machine not only missed the reference, but also renamed the sword slightly. Its new name, "*Kotatsu*", refers to something very different: a special Japanese table with a heat source built into it.

Shh! Be quiet in the library!

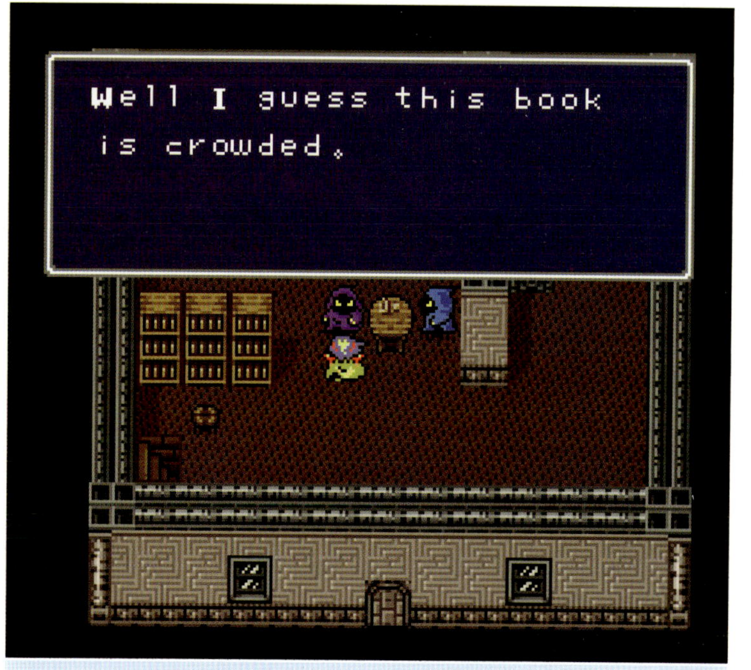

Wow, this book really moves me.

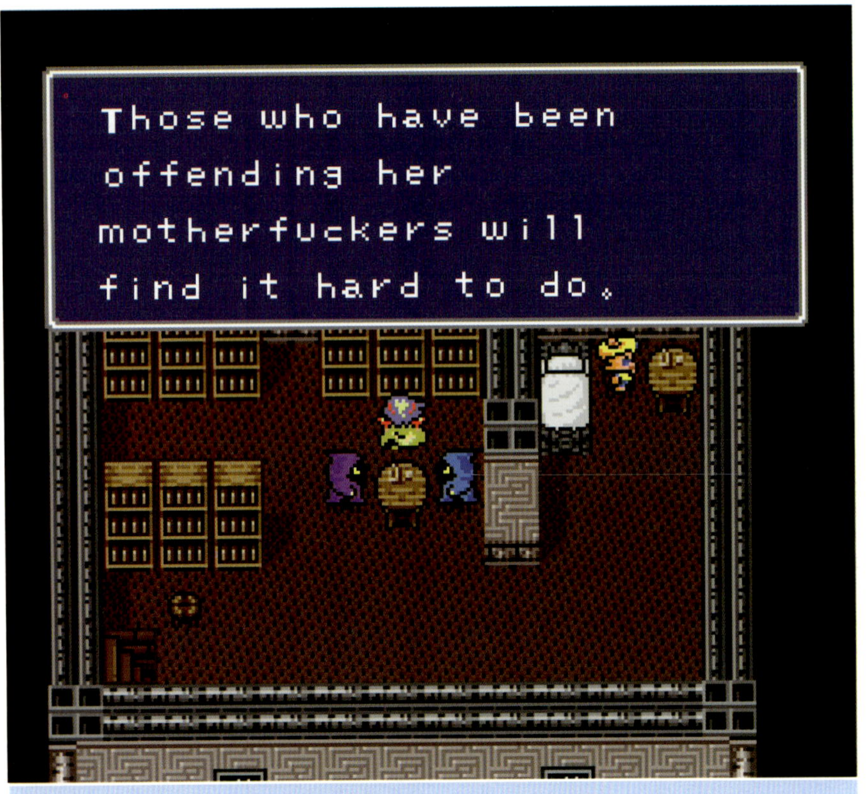

Those who anger her will be frozen to their very core.

◂ Swearing, Translation, and Machines

The Japanese language has its own share of rude and crude language, but it doesn't have the same kind of "swearing system" that's common in other languages. Instead of relying on specific offensive words, Japanese relies more heavily on shifts in politeness levels, intonation, and context clues to achieve the same results.

For example, there's indeed a crude Japanese word for excrement: *kuso*. Naturally, you probably wouldn't use the word around your boss, a random stranger, or a teacher – it'd be extremely rude and inappropriate. But *kuso* also appears in children's entertainment all the time without anyone batting an eye. So although *kuso* literally means "shit", whether or not it should be translated as "shit", "darn it", or something else depends entirely on the context.

Basically, it falls to the translator to decide when a Japanese word should be translated into a swear word or not. And since it's such a vague matter of interpretation, there's rarely a single "correct" way to handle crude Japanese. The result is that different translators will translate crude Japanese language very differently. Consequently, looking at how crude Japanese is handled in translation can give insight into a translator's mind.

But what happens when the translator is a machine? That's what's so fascinating about the swear words in these *Final Fantasy IV* machine translations: what's going on inside the machine's mind? From what we've seen, current machine translation systems have no problem using profanity and offensive words, nor do they take the concept of target audiences into account. As a result, Translation #2 not only features terrible translation accuracy, but plenty of offensive phrases as well.

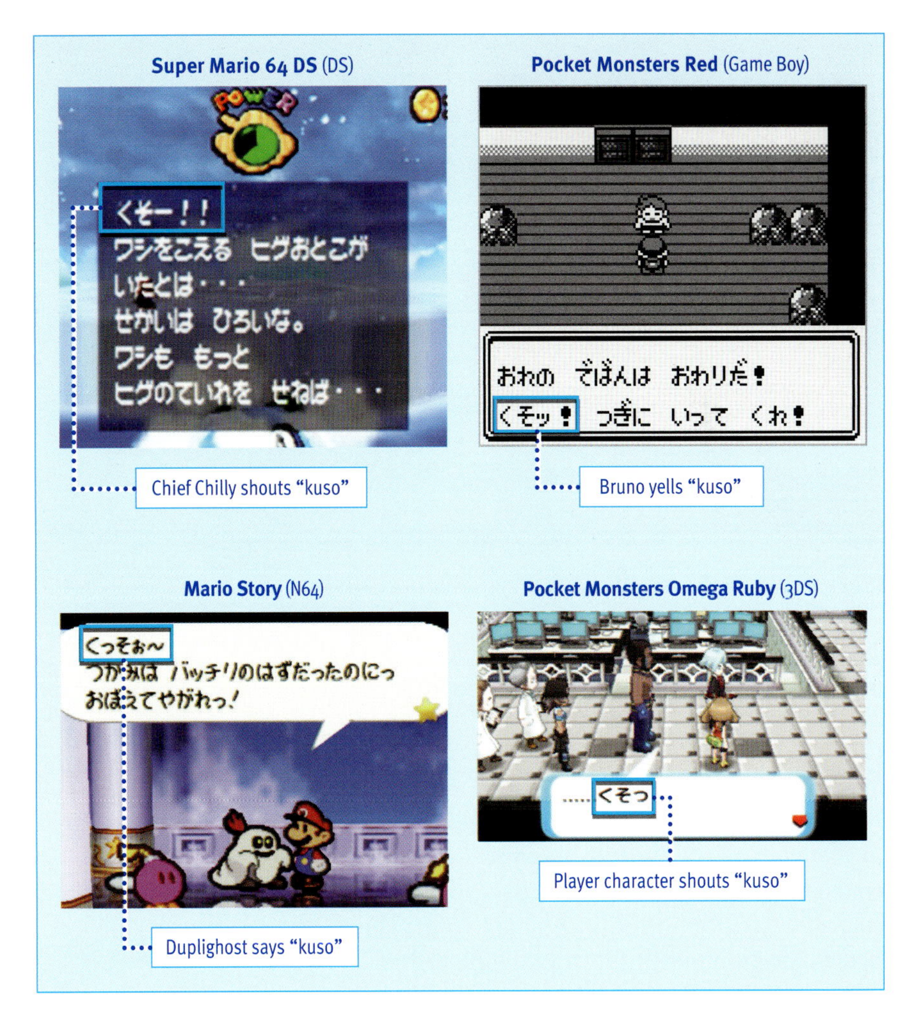

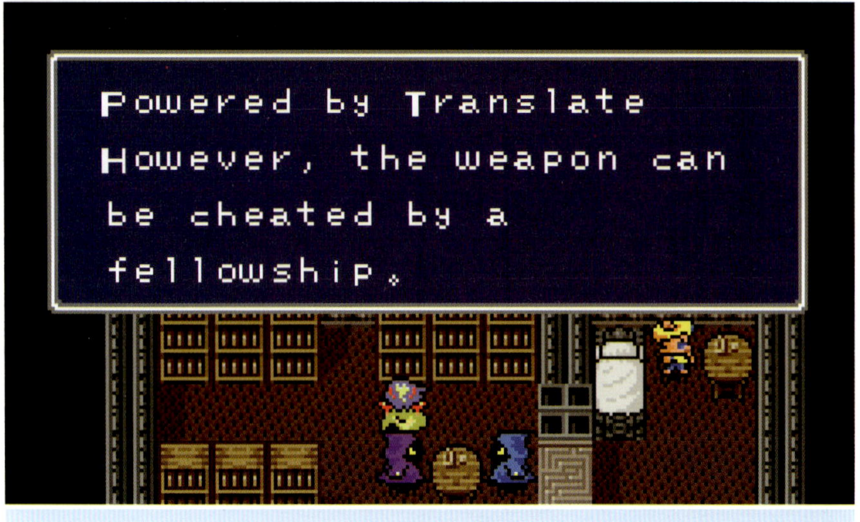

Ramuh is a mild-mannered old man… but his anger brings forth dark clouds and thunderbolts of punishment.

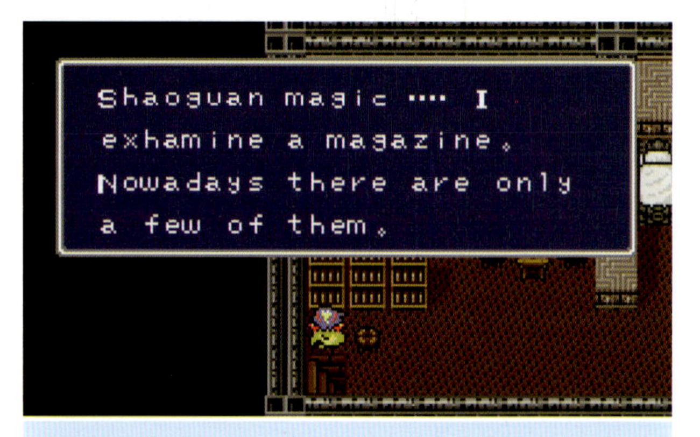

Summoning Spells… A type of magic that calls forth Eidolons. Very few with this power remain today.

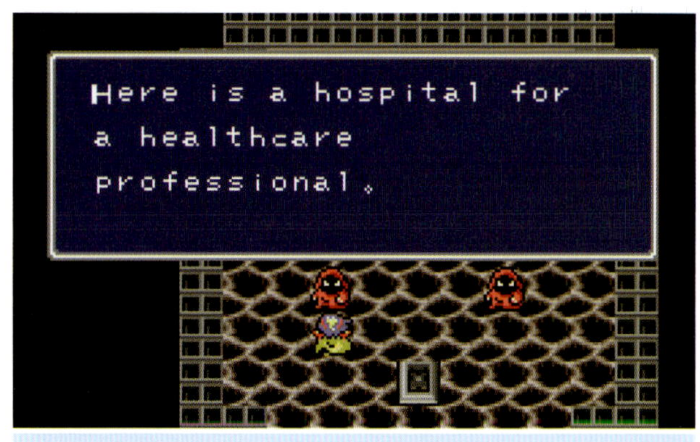

These are the chambers of the Eidolon King and Queen.

▲ A Friendly Reminder

Every so often, the phrase "Powered by Translate" appears in Translation #2. It seems to happen in the middle of large chunks of text, so I originally assumed it was a type of text "watermark" that Google Translate sometimes includes. After some experimenting, though, I discovered that this particular instance of "Powered by Translate" happens even if there's only one line of text being translated. During these tests, I found a few alternate translations of the first line, including "Warm thick lamb wax" and "Poisonous lasagna".

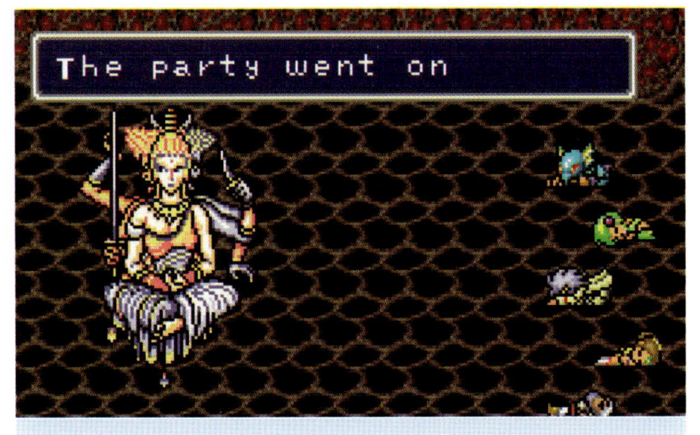

The party was wiped out...

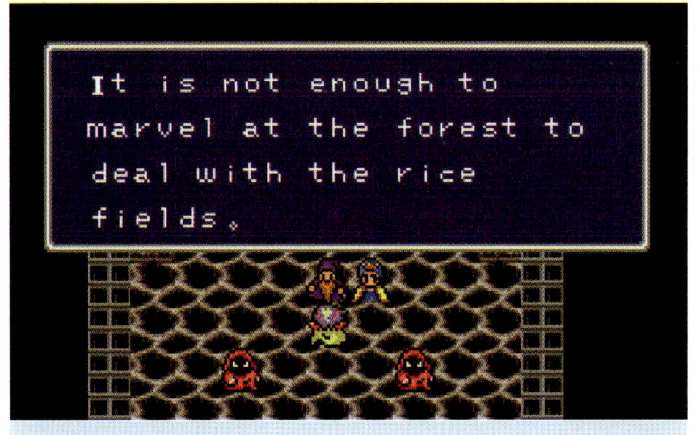

Great strength alone cannot defeat the power of evil.

Story

Golbez has all of the magical crystals, except for one. The Dwarf King sends the heroes to the Sealed Cave to protect the final crystal.

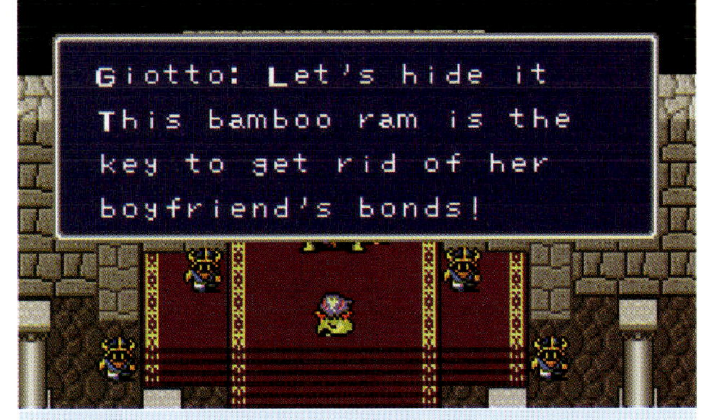

Giott: In fact, this necklace is the key to opening the Sealed Cave's entrance!

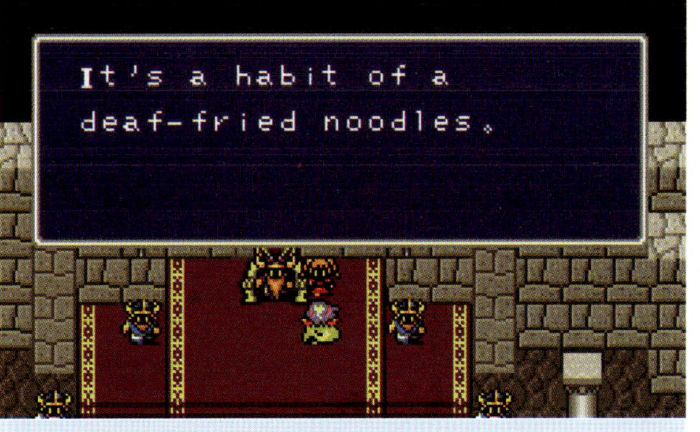

I'm told that necklace is a memento of my late mother.

Rumor has it there's a cave somewhere in the Underground that leads to the Eidolon Realm!

Cecil: The wall!
Rydia: It's moving!

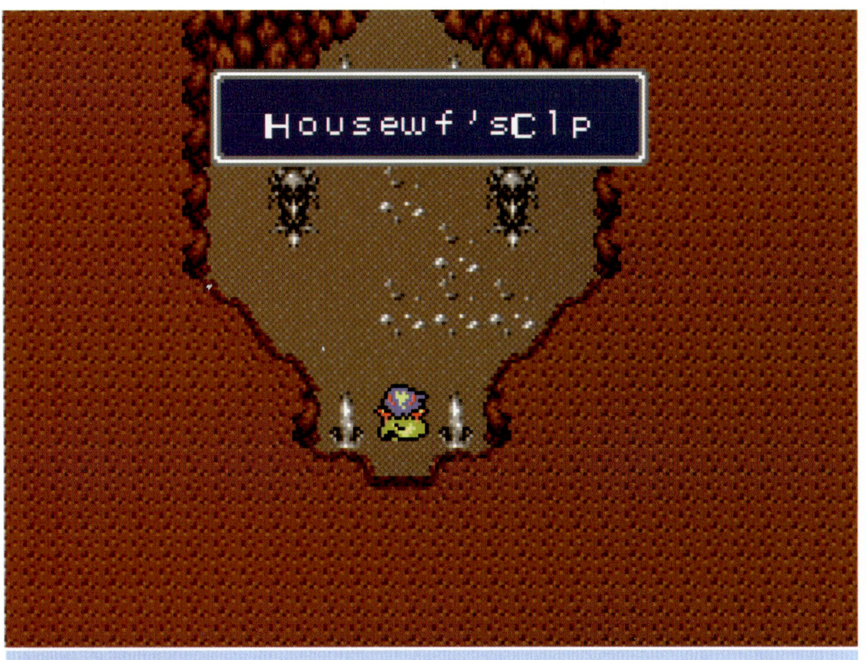

(Unshortened Text: "Housewife's clip")
Sealed Cave

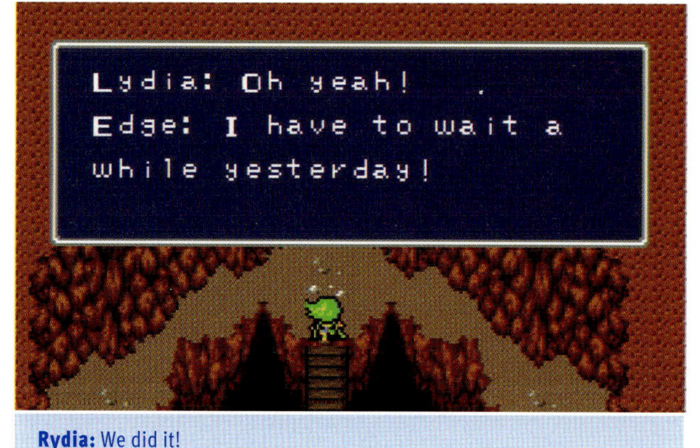

Rydia: We did it!
Edge: Better luck next time!

Cain: It's okay… I've come to my senses!

Story 

The heroes acquire the final crystal and defeat a demonic wall monster.

Just as the party reaches the cave's exit, Cain betrays Cecil, takes the crystal, and returns to Golbez's side.

With all of the crystals in hand, Golbez can reach the moon and obtain immense power.

▲ Past and Present Collide

In this scene, Edge uses a stock Japanese taunt that literally means "Bring it on the day before yesterday!" Although this literal translation makes some sense in English, a skilled human translator would recognize that it would be better to use an equivalent English taunt instead. Machine translators don't appear to have that same level of intuition or creativity at the moment.

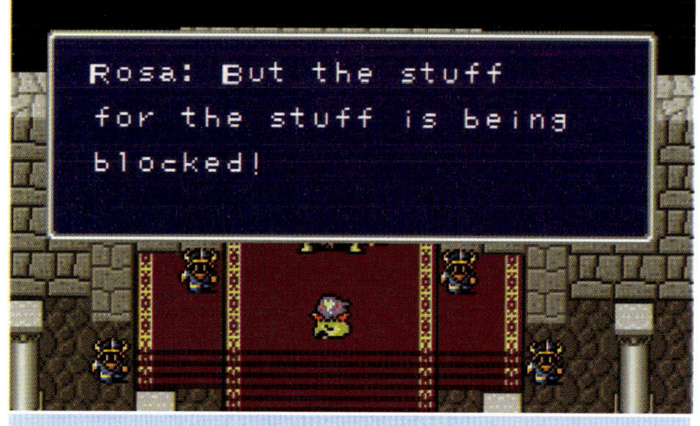

Rosa: But the path to the surface is blocked!

Rosa: Are your injuries all better now?

◂ Ready When You Are

As we've seen, context lies at the heart of the Japanese language – a single, everyday sentence can have a dozen different meanings depending on the situation. Part of the Japanese-to-English translation process is identifying these different meanings, and then choosing the proper meaning based on all available information.

This line from Translation #2 sounds very wrong at first, but it's technically a valid translation of the original Japanese phrase. A human translator, even without knowing the game's plot, would recognize that "Are you ready for scratches?" is much less likely to be the intended meaning than something like "Are your injuries all better?". This is a nice example of how translation is rarely a clear-cut affair – intuition and personal experience play a part in the translation decision process.

The most I can do is fiddle with your airship...

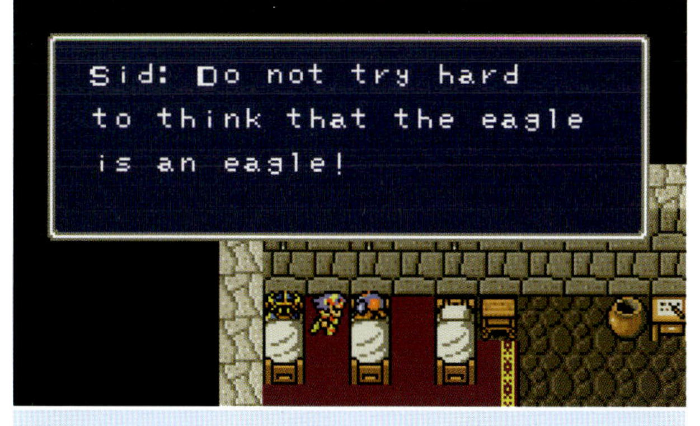

Cid: Think of the airship as my double and go do what you gotta do!

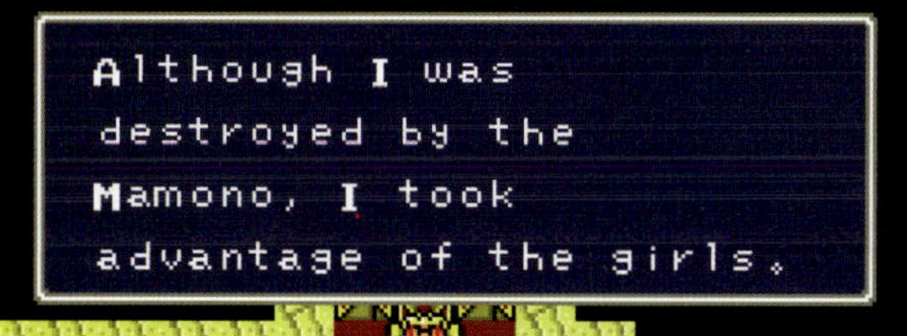

Cecil: Your Majesty... **King of Baron:** Do not look so sad.
Although I was slain by a monster, I also gained everlasting strength.

Story

More side quests become available.

Several quests involve finding, defeating, and befriending powerful Eidolons.

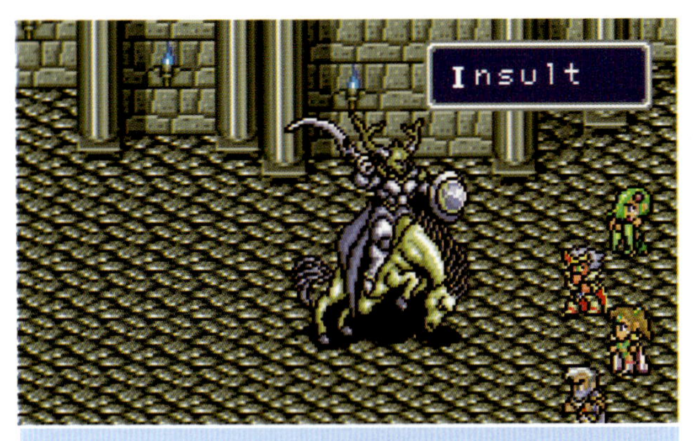

Zantetsuken

World of Final Fantasy (PS4)

◀ Cut It Out

A character named Odin has a deadly attack that can kill multiple foes in a single hit. In Japanese, this attack is known as *Zantetsuken* ("iron-cutting sword"), a reference to a type of Japanese sword famous for cutting through iron swords and armor. The *Zantetsuken* name is so famous that it appears regularly in popular Japanese fiction, including the *Final Fantasy* series. It's such a staple of the *Final Fantasy* series, in fact, that it's been left as "Zantetsuken" in most English translations.

Google's machine translator had no knowledge of this historical connection or how the term has been translated in other *Final Fantasy* games. Ironically, a human translator probably would've used an online search engine to look up the name and learn these background details. Perhaps machine translation systems of the future will also be able to connect with search engines for additional information.

Still, why did the machine translate *Zantetsuken* as "Insult"? I couldn't see a clear connection myself, so I tried experimenting with the term. I didn't find anything conclusive, but I did encounter alternate translations like "Knotted sword", "Killing", and "Ironworks".

Real-Life *Zantetsuken*

The electric shock ripped through Odin!

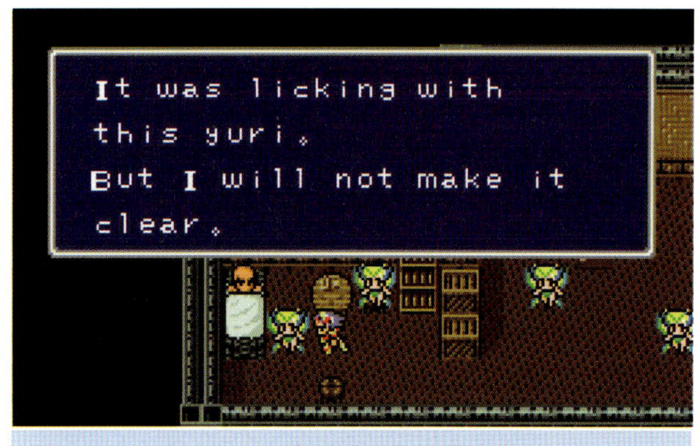

He was lying on the ground outside the cave. He hasn't woken up at all, though.

Honestly, the nerve of that man, lying around all day like that...
Oops, I must've gotten some dust in my eyes.

Obtained Kitchen Knife!

Story 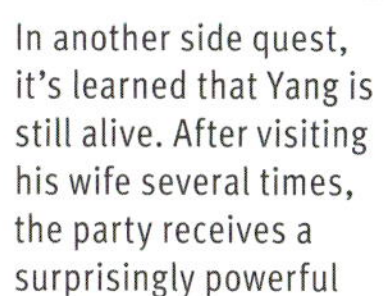

In another side quest, it's learned that Yang is still alive. After visiting his wife several times, the party receives a surprisingly powerful weapon.

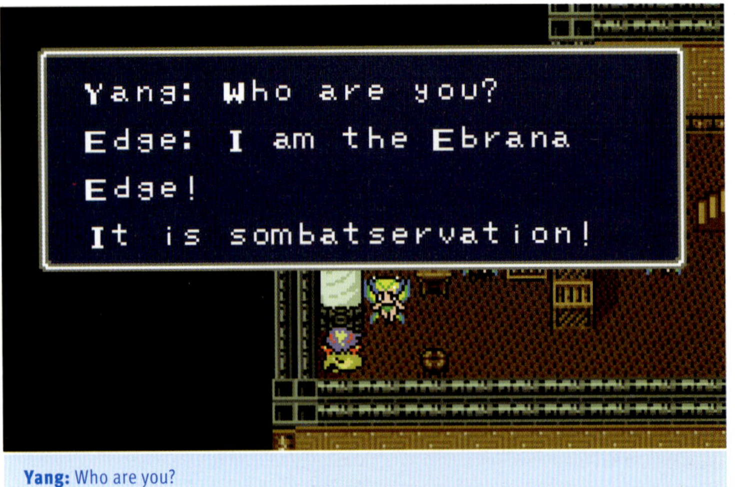

Yang: Who are you?
Edge: I'm Edge of Eblana! I'm even more skilled than you!

▲ English 3.0

As we've seen, the machine responsible for Translation #2 occasionally invents brand-new English instead of translating text into existing English words. This line, which features the word "sombatservation", is my favorite example of this unusual behavior.

When I first saw "sombatservation" in Translation #2, I assumed it was a fancy, obscure English word that the machine pulled from the depths of the language. Later, when I tried to look up what it meant, I discovered that it's *not* an English word, nor has it ever been. So where did it come from?

After experimenting with the Japanese sentence in question, I could find no clear cause for "sombatservation". We've seen how the machine has difficulty with unusual, entertainment-style Japanese dialogue, but this particular text is still relatively close to standard Japanese. Even so, the machine translator is very touchy with this sentence – simply changing the exclamation mark at the end produces entirely different translations, such as "Sorry for bothering you" and "It is saddened by you".

I also tried making the sentence easier for the machine to parse by adding spaces and extra formatting, but it only made things worse. Some examples include:

- The fun of being fried on you!
- Tomatasu you on a cold day!
- Tatashi warmth over you!
- You will feel better by all means!
- You taught me!
- Trout played in a cup of tea!

Out of curiosity, I entered "sombatservation" into the machine translator and had it translate it back into Japanese. In December 2016 it translated it as "gymnasium", but as of October 2017 it translates it as "meteorological observation".

▶ Cruise Control for Cool

As we've previously covered, sometimes an English translation might be presented in all caps if the Japanese text is written entirely in *katakana*. This stylistic choice is common among human translators, but it's rare to see it from machine translation systems.

In this particular line, the Japanese name for "Mysidia" – which is always written in *katakana* – was replaced with the name "Michitie" and presented in all-caps.

Mysidia
written in *katakana*

This use of all caps comes as a surprise, not only because it's rare to see machines follow this translation style, but because the name was translated as the normal-looking "Missidia" throughout most of Translation #2. What caused the machine to treat this instance of "Mysidia" differently? At the very least, we can see that it has some rudimentary knowledge of different presentation styles.

According to Mysidia's records, there's a flight crystal separate from the standard flight controls. Supposedly, if you talk to it, it will take you to and from the moon!
Cecil: I'll give it a shot!

Story 

The heroes visit Mysidia. The village elder recites the local legend, which speaks of a flying ship that can go to the moon.

The Mysidians pray for the legend to come true. A weird spaceship shaped like a whale rises out of the ocean.

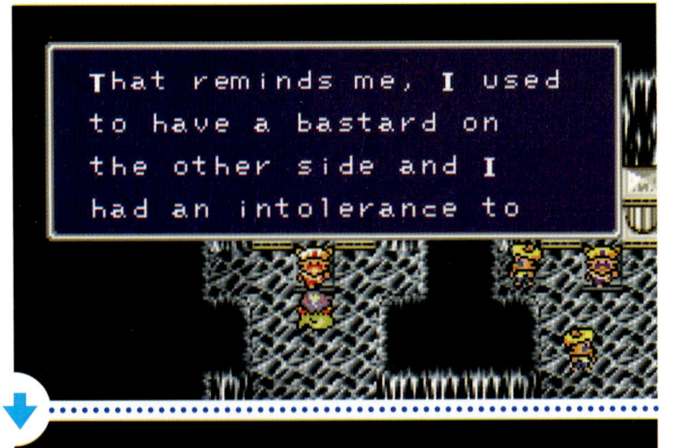

Come to think of it, one of our friends invented a ship and took it to the Blue World. He was really good at coming up with nicknames too...

We're the Hummingway clan.

Story

The heroes fly to the moon and meet Fusoya, a Lunarian. Before he joins the party, he shares important plot details.

It turns out that Golbez and Cecil are actually brothers, and their father is a Lunarian. Golbez has been brainwashed by a Lunarian named Zemus, the true villain of the game.

▲ Happy Little Accidents

Every once in a while, the stars align during the translation process, resulting in a translated line that accidentally transcends the original text. Most human translators will probably experience this at one point or another – maybe a translated joke turns out to be wittier than normal, or maybe a translated phrase sounds so good it unintentionally becomes a memorable quote.

I never stopped to consider if this phenomenon could happen with machine translators too, until I came across this simple example. Although the machine translation is clearly wrong, its accidental "a way away" phrase has a bit of sing-songiness that surprisingly fits this area's musical theme.

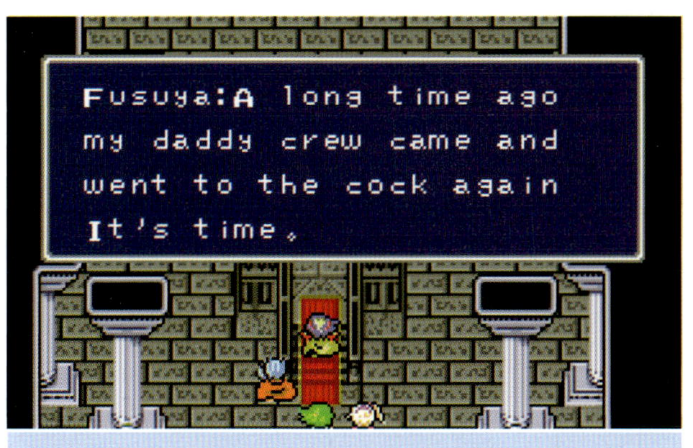

Fusoya: My younger brother Kluya built it and took it down to the Blue World long ago.

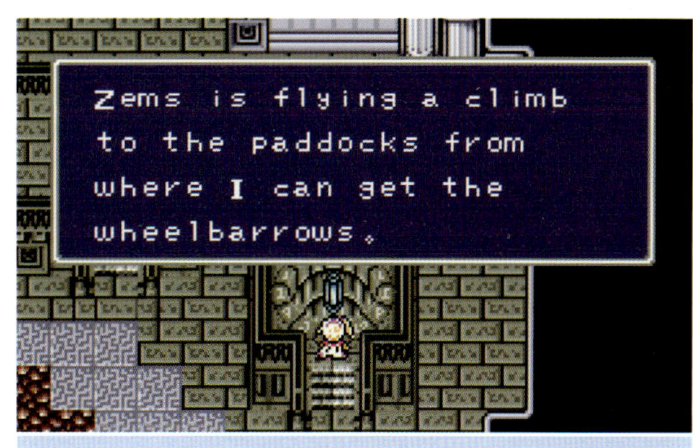

Zemus has created a barrier from the inside to block the path leading to the core.

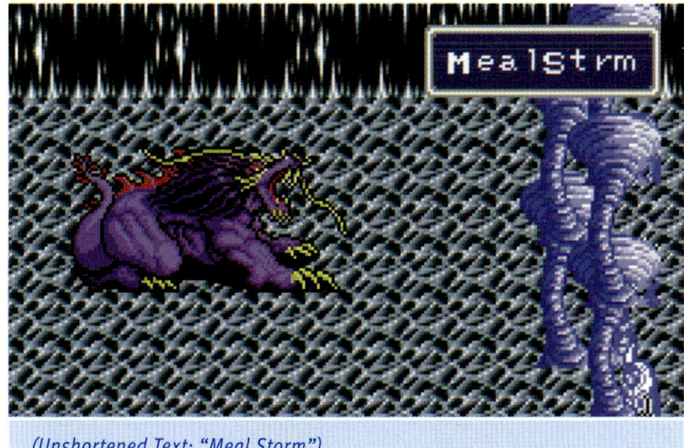

(Unshortened Text: "Meal Storm")
Maelstrom

So, you've made an ally of Leviathan.

Story

The heroes explore a cave on the moon. Deep inside, they fight and befriend the ultimate Eidolon: Bahamut.

Mato Says

Fusoya's name is a little closer to "Fusuya" in Japanese, which is why he's called "Fusuya" in Translation #2.

For simplicity's sake, we'll stick with "Fusoya", the English name used in every official English release of *Final Fantasy IV*.

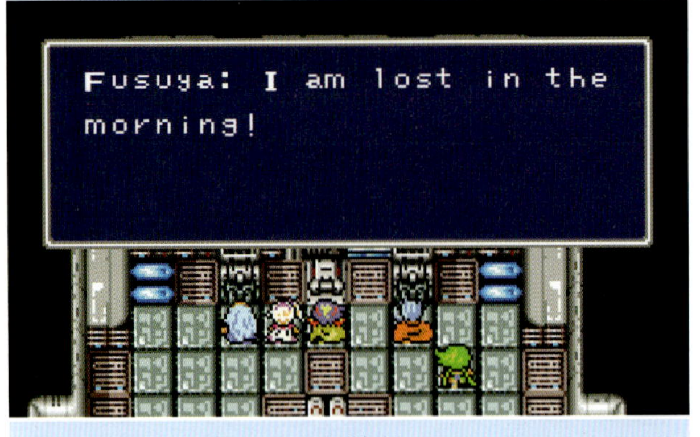

Fusoya: Now is our chance to get inside it!

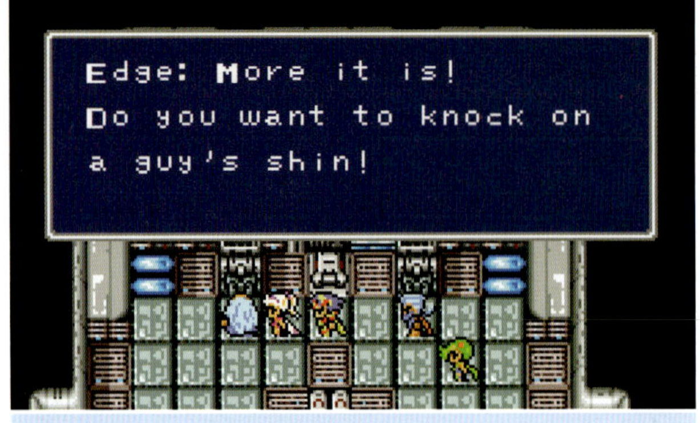
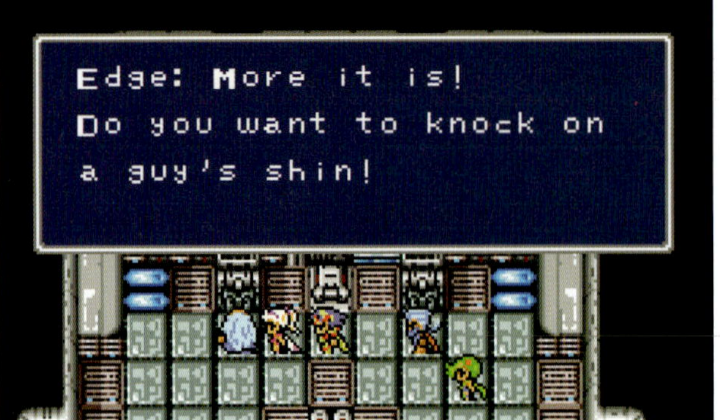

Edge: Aha! So we're gonna bash the thing's heart!

Fusoya: Now!
Cid: Hold on tight, guys!

Giant's Neck

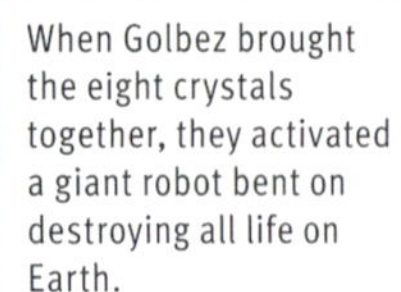

Story

When Golbez brought the eight crystals together, they activated a giant robot bent on destroying all life on Earth.

The people of Earth come together and distract the robot long enough for Cecil's party to sneak inside it.

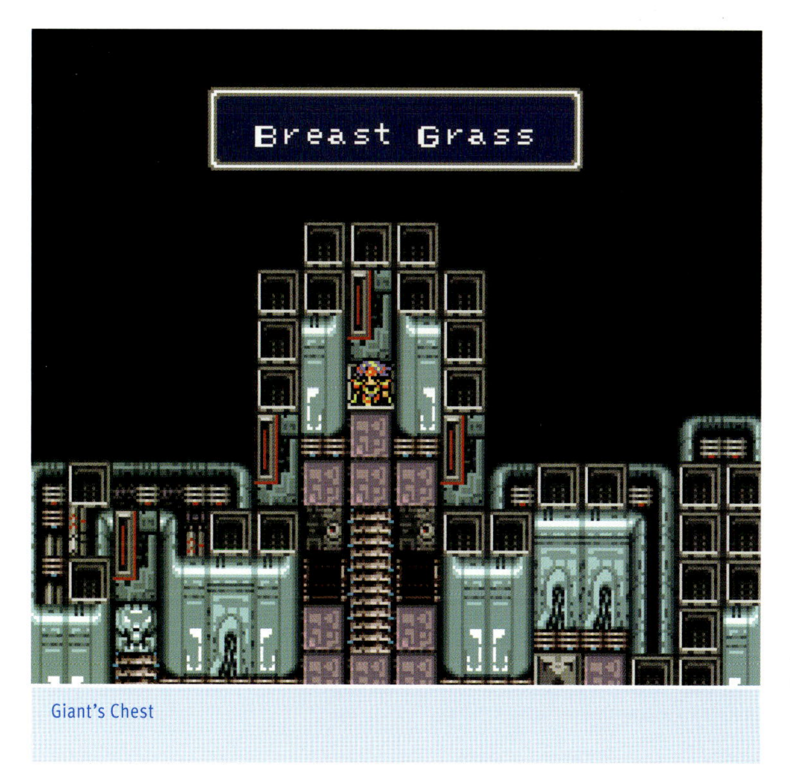

Giant's Chest

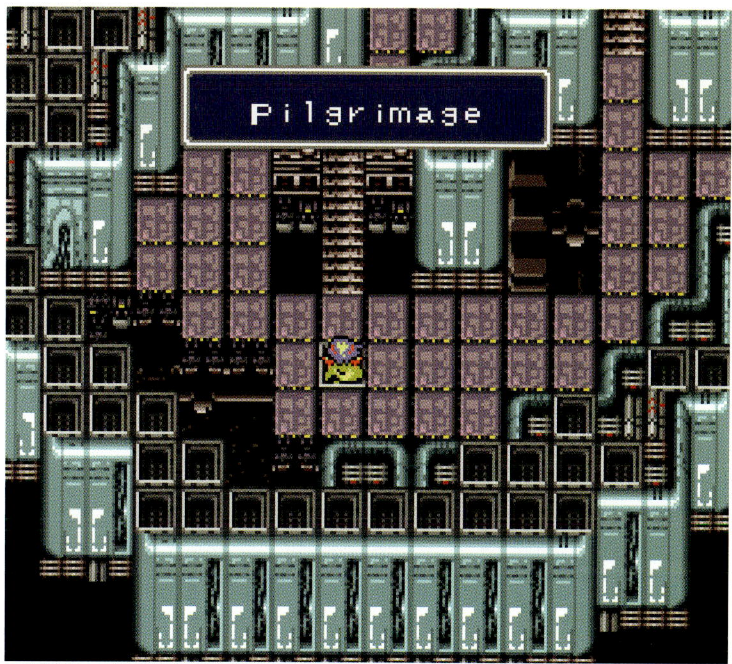

Giant's Stomach

Giant's Inner Corridor

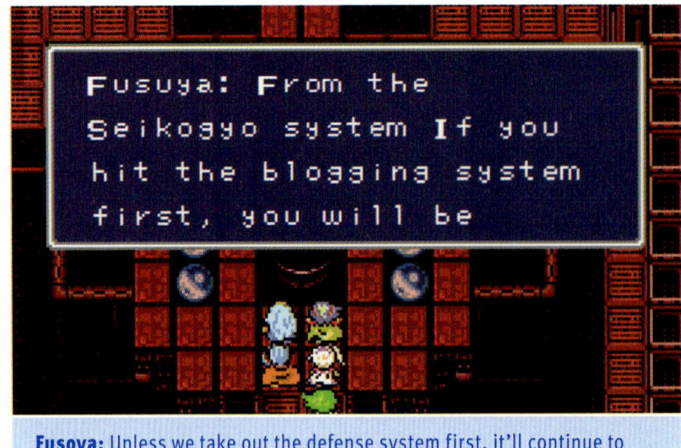

Fusoya: Unless we take out the defense system first, it'll continue to heal the control system!

Story

The heroes eventually reach the giant robot's control center. After they destroy its core, Cain and Golbez break free of Zemus' control.

(Unshortened Text: "Subjugation tobacco")
Giant's Inner Stairwell

▲ Wounding WordPress

In this obvious mistranslation, the Japanese word *bōei* ("defense") was replaced with the word *burogu* ("blog") before it was translated. This pattern of unrelated word replacement is nothing new in Translation #2, but it's possible the discussion of multiple "systems" in this line prompted the machine to use a modern computing term.

Fusoya: This is the control system in the Giant's heart!
Edge: It's huge!

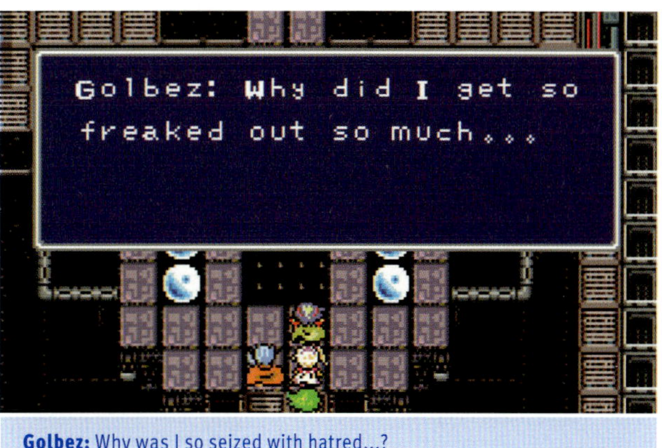

Golbez: Why was I so seized with hatred...?

Cecil: Still, it could've been just the opposite too...

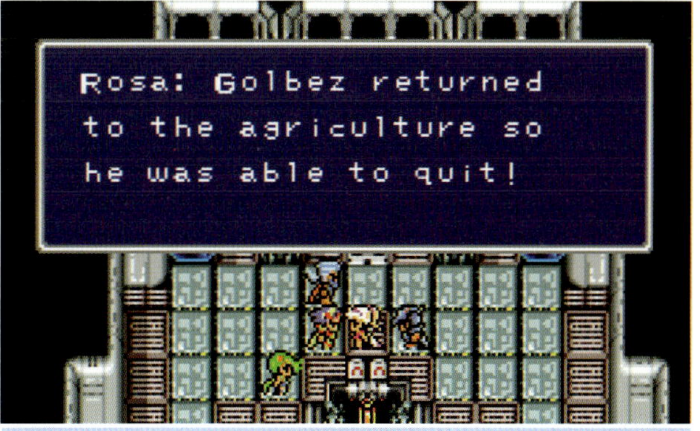

Rosa: Golbez has returned to his senses too! The spell must've been broken!

Story 

Fusoya and Golbez head to the moon to face Zemus.

Cain rejoins Cecil's party, but feels great remorse for his actions.

Reader Challenge

Quick! Think of two English synonyms of the phrase "freaked out"!

Cain: ...If that happens, do not hesitate to kill me!

▲ Wanderers from Es

In this emotional scene, Cain uses the word *kiru* ("to cut", "to kill"). Instead of translating *kiru*, however, the machine first morphed it into *kiku* ("to listen", "to hear").

So how does "erroneously" fit into the translation? After some testing, it seems the machine took the Japanese phrase *enryonaku* ("without hesitation"), saw the "e" at the start, and chose an English word that also starts with "e". I encountered several other translations of *enryonaku* during my tests too, including "enjoy", "enhanced", and "encouragement". Whatever the machine is doing here, it isn't translation.

Edge: Okay, now be a good girl and stay put while we're gone.
Rydia: You idiot!

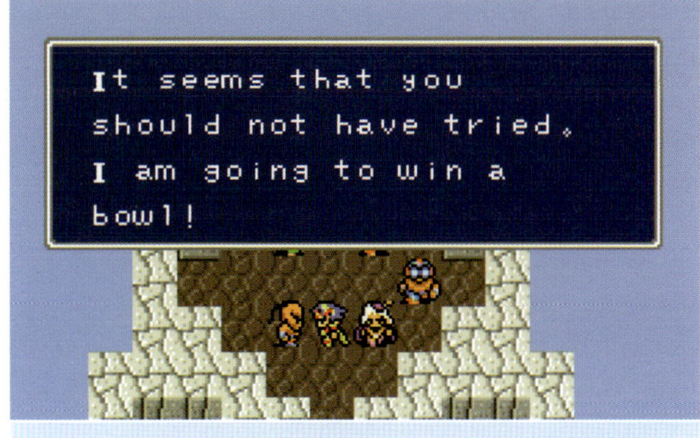

It seems it was destined for you to do battle with him. I'm praying for your success in battle!

Why are you banging
with an eagle!
Not at all I do not
want to go live!

Why won't you take me with you?! Sheesh... Listen up! You better come back alive, you hear?!

Story

After a bit of drama, the heroes make their final preparations to return to the moon and battle Zemus.

▲ Lucky Dishes

The Japanese word *bu un* ("good fortune in battle", "good luck") appears regularly in *Final Fantasy IV*. Unfortunately, the machine translator struggled with the term so much that it consistently translated it as "bowl" instead. As a result, friends and well-wishers throughout the game seem more obsessed with dishware than the fate of the world.

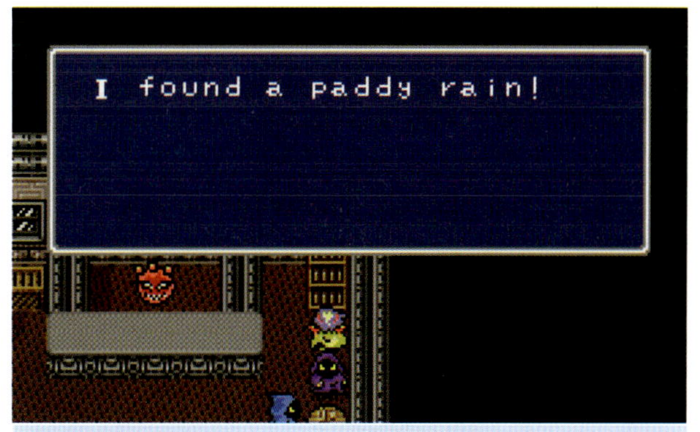

Found Secret Technique Book!

Mato Says

Not many fans know this, but after you defeat Bahamut, this hidden item becomes accessible back in the Eidolon Realm. It doesn't work in the original Super NES translation, though. Don't tell anyone I told you this – it's a secret to everybody!

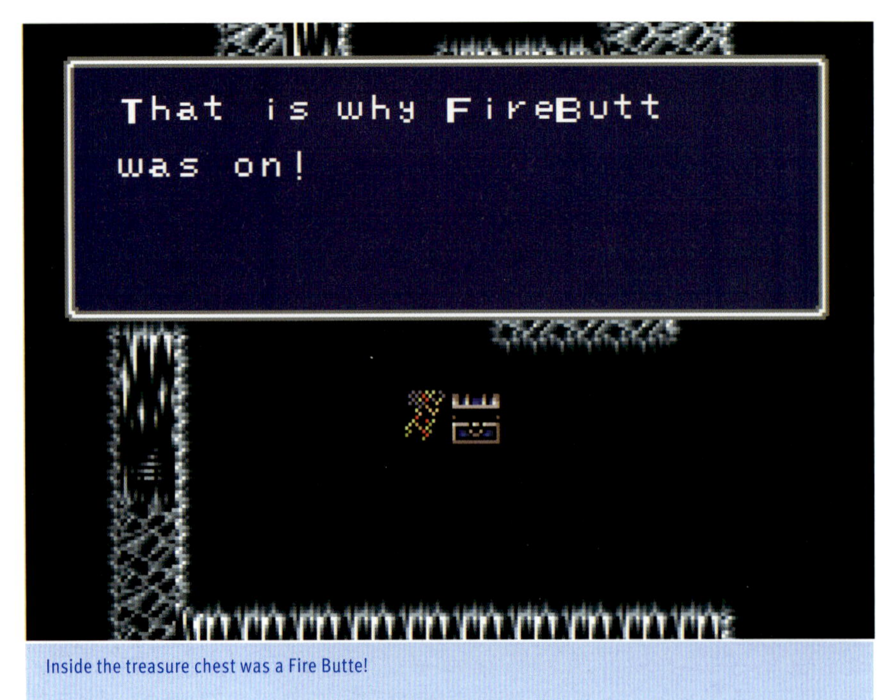

Inside the treasure chest was a Fire Butte!

◂ Getting Cheeky

Item names can only be eight letters long in *Final Fantasy IV*, so my automatic text compressor had to trim most of the machine-translated item names until they fit. Sometimes these abbreviations worked well, and other times they became incomprehensible.

In this instance, the "Fire Butte" whip had its name shortened until it became the "FireButt" whip. Funny name abbreviations go hand-in-hand with classic video games, so this silly name is actually in line with old translation practices.

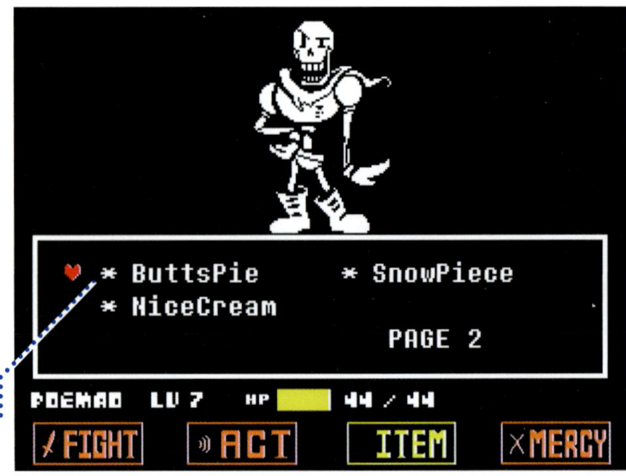

Undertale (PC)

Bizarre text abbreviations were such a big part of the early gaming experience that some modern games use abbreviated text to achieve a retro vibe. In this example, the "Butterscotch Pie" item name was intentionally abbreviated for comedic effect.

Saurus Zombie

Heat Ray

Story

The heroes enter the final area of the game. They travel deeper and deeper underground until they reach the lunar core, where Zemus awaits.

▲ Spooky Sauce

Without any context to work with, the machine translator had to decide between calling these enemies "Saurus Zombies" and "Saulus Zombies". It made a wild guess and chose the latter spelling. To make matters worse, my automatic name compressor dropped some key letters when shortening the name. The end result: these terrifying skeleton monsters now look like they're called "Salsa Zombies".

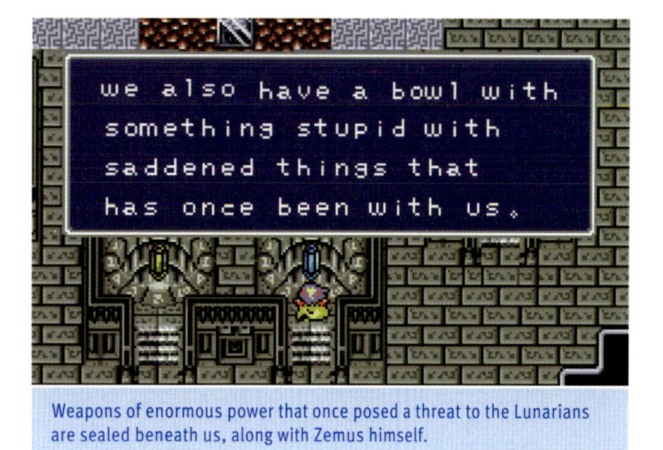

Weapons of enormous power that once posed a threat to the Lunarians are sealed beneath us, along with Zemus himself.

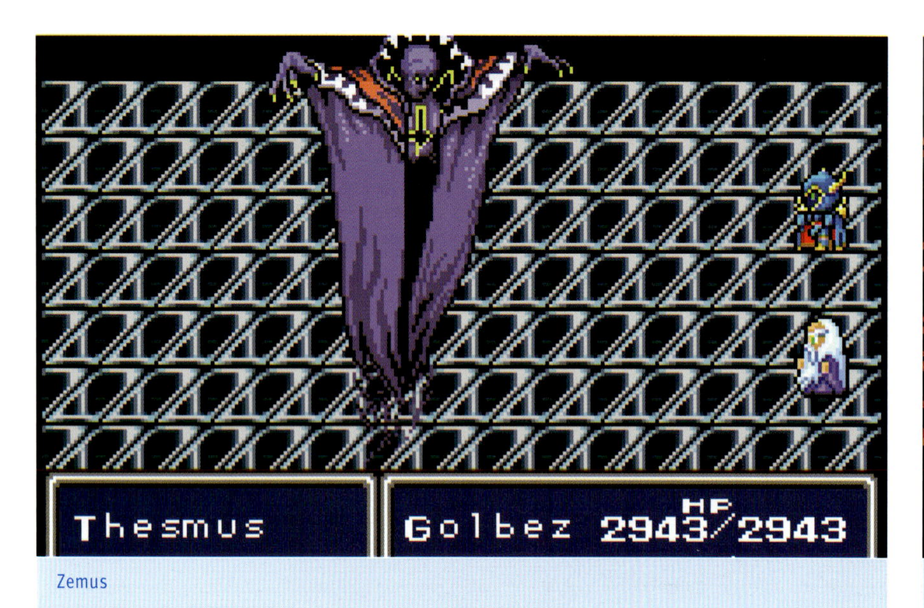

Zemus

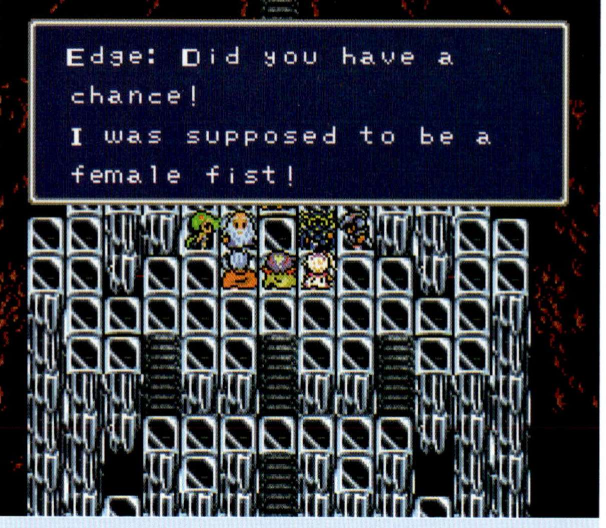

Edge: We were one step too late! I was supposed to kick his butt!

The Many Names of **Zemus**

Zemusu	Zemu	Themes
Zemus	Zems	Thesmus
Zemuz	Zeus	Theemus

▲ So Many Forms, So Little Time

The neural network translation system behind Translation #2 is easily described with one word: indecisive. This is especially clear near the end of *Final Fantasy IV*, when the heroes encounter the game's final villain: Zemus. During the final act of the game, Zemus is referred to by many different names. What's more, most of these names appear in the span of just a few minutes. His name transforms so often that you can almost "feel" the machine's struggle to decide on a name!

◀ Punch Line

The biggest mistake in this line – the "female fist" part – follows the same mistranslation patterns we've seen before, but with an extra step added. In fact, it's a mistranslation of a mistranslation of a misreading.

First, the line in question features the Japanese word *buchinomesu* ("to thoroughly beat up", "to kick the crap out of someone"). For some reason, this common word was replaced with the phrase *bushi no mesu*, which was then further changed into *kobushi no mesu*. The word *kobushi* means "fist", and *mesu* means female, so the entire phrase might translate as "the fist's female" or "the female of the fist". The final translation flipped these words to reach "female fist", arguably making it a mistranslation of the mistranslation it was based on.

From a translator's point of view, the convoluted process behind this mistranslation is almost as amusing as the final mistranslation itself!

I... am sheer dark matter...
The amassment of Zemus' hatred...

▲ Transla-ja Vu

This ridiculous mistranslation is a perfect example of how the fancy machine translator responsible for Translation #2 sometimes "echoes" itself, resulting in repeated words and phrases. This example also demonstrates how words within these echoes tend to share the same first letter. I'm not sure why this is, but I assume that it's related to all the other mistranslation patterns in Translation #2 that involve the first letters of words.

I am Zero Mus. I will
cry all to ····!!

Cecil: ZERO MUSIC!
It will not be okay...

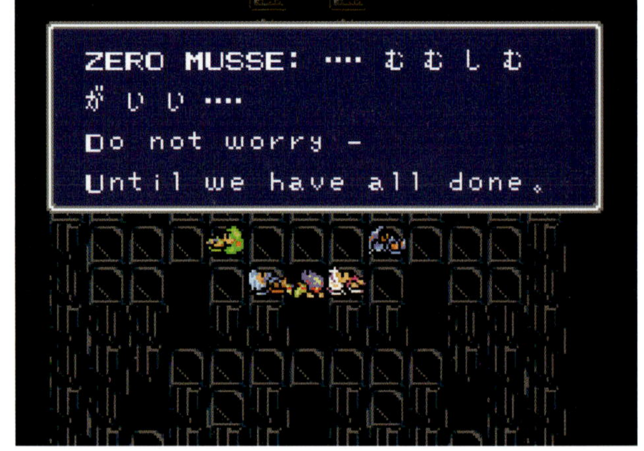
ZERO MUSSE: ···· むむしむ
がいい····
Do not worry -
Until we have all done.

◀ Master of Name Changes

Just before the final battle begins, the main villain transforms from a man named Zemus into a horrifying creature known as Zeromus. By sheer coincidence, the machine translator kept this transformation theme alive by giving Zeromus a barrage of new names, most of which appear in the span of about 60 seconds.

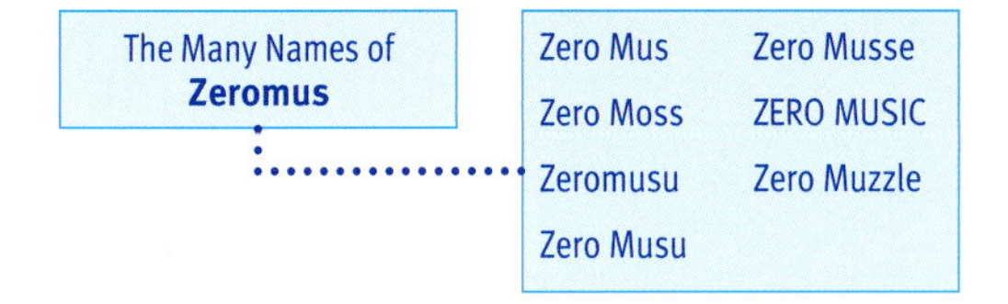

The Many Names of
Zeromus
Zero Mus
Zero Musse
Zero Moss
ZERO MUSIC
Zeromusu
Zero Muzzle
Zero Musu

Because the name "Zeromus" only appears a handful of times in the game, I was curious to see if this cutting-edge machine translator would continue to create new variations of the name. To test this out, I fed the machine simple Japanese phrases that included Zeromus' Japanese name. I was immediately greeted by a dozen more name deviations, including "Zero Mouth", "zero musk", "zero mill", "Zero Muscle", and "zero moth".

Story

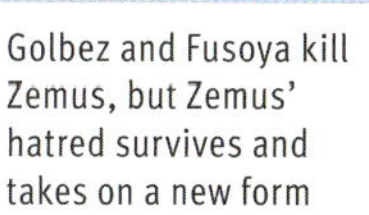

Golbez and Fusoya kill Zemus, but Zemus' hatred survives and takes on a new form known as "Zeromus".

Zeromus instantly wipes out everyone, including Cecil and his friends.

Golbez and Fusoya struggle against Zeromus, but the darkness in Golbez's heart makes it a futile effort. All hope seems lost.

Reader Challenge 

Quick! We can't let the machines one-up us! Can you think of 5 other silly names for Zeromus that start with "Z" and "M"?

Palom: Hang in there, dude!

The prayer granted them the power to endure!

Story 

Everyone in Mysidia starts to pray for Cecil and his friends. The prayers revive the heroes and give them the strength to fight Zeromus.

Mato Says 

The "Shin-chan" line is especially amusing to me because I have fond memories of working on the show's official translation years ago. So seeing my different interests collide like this was a nice surprise!

▲ To Be a Man, You Must Have Honor

In this mistranslated line, the word *anchan* ("big brother" / an informal version of "mister") was replaced with the name "Shin-chan", presumably because of their similar spellings. This is just my best guess, though – it's hard to tell what the machine was actually thinking here.

Incidentally, "Shin-chan" is the name of a popular Japanese cartoon character often described as a Japanese Bart Simpson. Because Shin-chan is so well known around the world, it's not too surprising that the machine translator would be familiar with the name.

▶ It's Just a Theory

The final boss' ultimate cosmic attack is called *biggu bān*, a Japanese pronunciation of the English term "Big Bang".

Unfortunately, when English word pronunciations are converted into Japanese pronunciations, some sounds and letters are dropped. As a result, very different English words sound the same in Japanese. For example, "soccer" and "sucker" share the same pronunciation of *sakkā*. Usually the intended meaning is clear based on context or intuition – you probably wouldn't talk about kicking a "sucker ball" into a goal, for instance. In this case, *bān* was understandably misinterpreted as "Burn".

The machine isn't entirely at fault for this "Big Burn" mistake, though – the game developer bears some of the blame. Normally, "Big Bang" is spelled *biggu ban* in Japanese, but the writers chose the atypical *biggu bān* instead. The elongated "a" pushes the word closer to "Burn" than "Bang", but context pulls it back to "Bang".

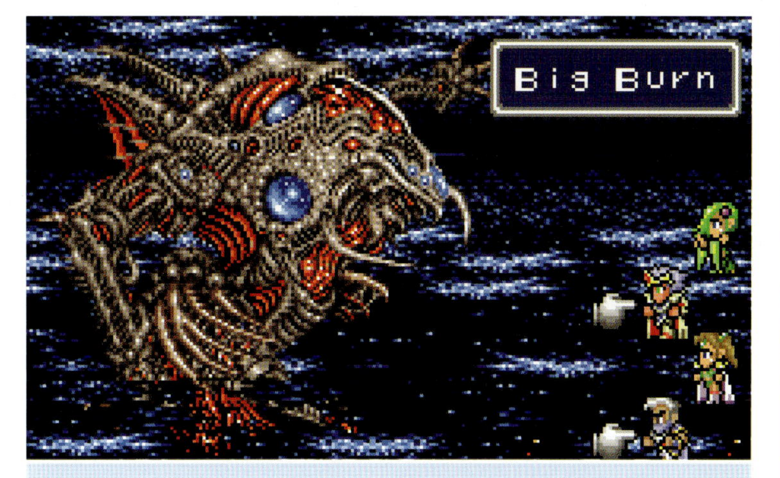

Big Bang

Final Fantasy V (PS1)

A real-life example of confusion between "Soccer" and "Sucker" during translation

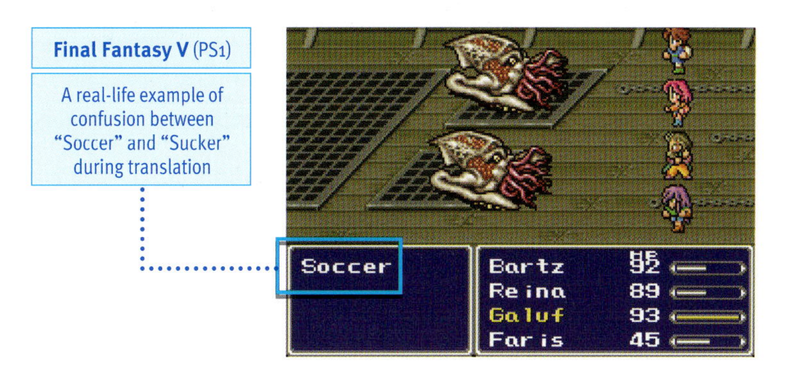

Story 

The big final battle begins. Zeromus unleashes one powerful attack after another.

The heroes persevere and finally defeat Zeromus.

Mato Says 

The Zeromus character in *Final Fantasy XII* uses a Big Bang attack too, and they fixed the Japanese name to use the less confusing *biggu ban* spelling.

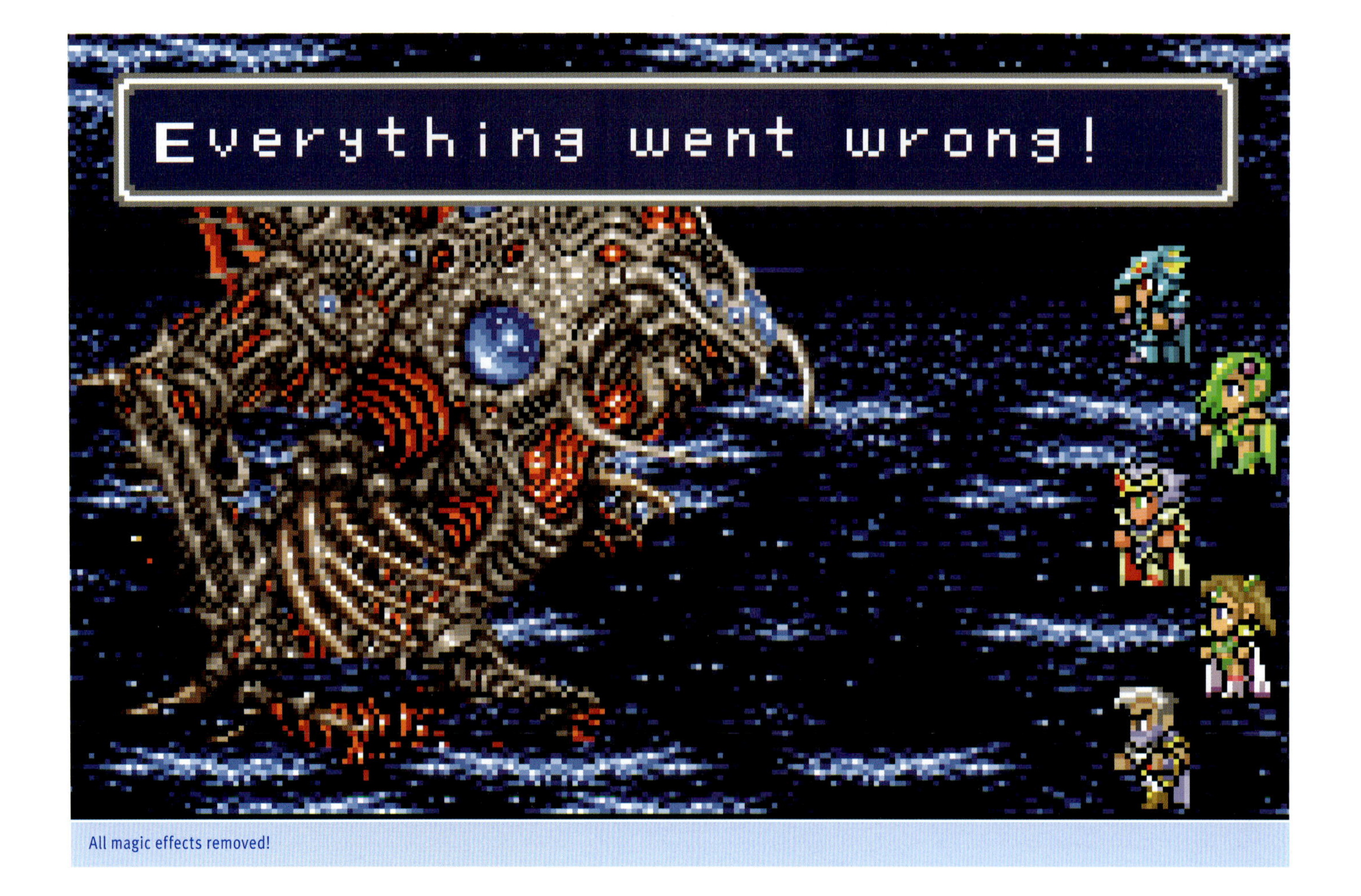

All magic effects removed!

Edge: Heheh! I do love justice, after all!

Fusuya: Really? I got a wonderful baby.

Fusoya: I see. You've gained wonderful friends.

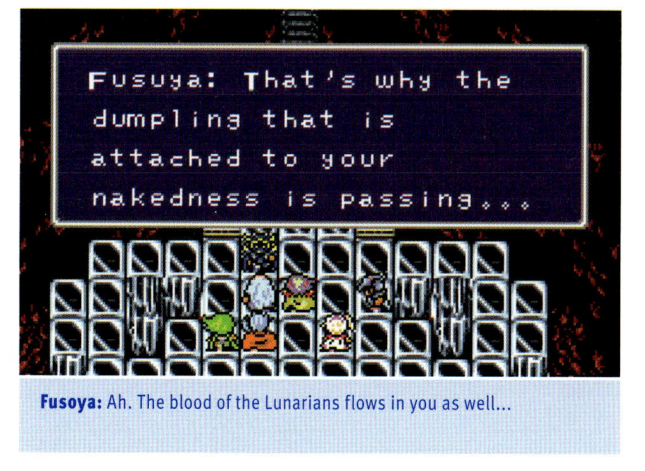

Fusoya: Ah. The blood of the Lunarians flows in you as well...

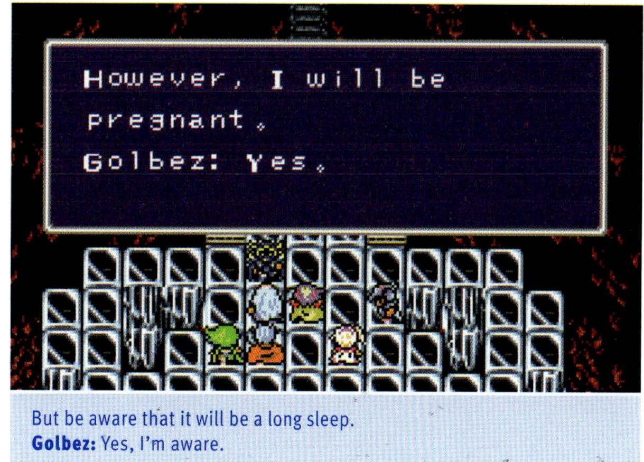

But be aware that it will be a long sleep.
Golbez: Yes, I'm aware.

Story

Golbez and Fusoya decide to remain on the moon in in a type of cryogenic sleep.

Cecil and Golbez acknowledge each other as brothers and share a first – and final – farewell.

Machine Translation	Intended Translation
However, it is a frozen peculiarity	It will be but a short reprieve
For each one asking for his or her light	The moon, in search of its own light,
I will not be lazy every time	shall depart on a new journey
A quarrel with a single girlfriend	One of the same blood shall remain on the moon
One is to become a garbage	One shall remain on the Blue World
The fate of time does not change itself	And the flow of time shall draw them apart

▲ Mistranslation: The Legend Continues

Mysidia's ancient legend reappears during a special ending scene, followed by new text that completes the legend. As before, the legend's "ye olden times" style of writing causes the machine to fumble through the entire thing. In the end, almost no trace of the original legend remains in the machine translation.

Story

The story comes to a close. Cecil and his companions return to Earth as great heroes.

The people of Earth work hard to repair the damage that was done. Characters get married, take the throne, and find true happiness.

The future looks bright.

Old Man: Young Master! You lack any sense of responsibility as the successor to the throne!

▲ Maybe Nobody Will Notice

The machine-translated version of this line is remarkably shorter than the Japanese line. Where did all the text go? The cause is unclear, but my best guess is that the machine had trouble deciphering the Japanese line and decided to discard a large portion of it.

It's also unclear how the old man came to be known as "REALLY" – the translation doesn't follow any previously observed patterns. In fact, the entire translation is REALLY a mystery.

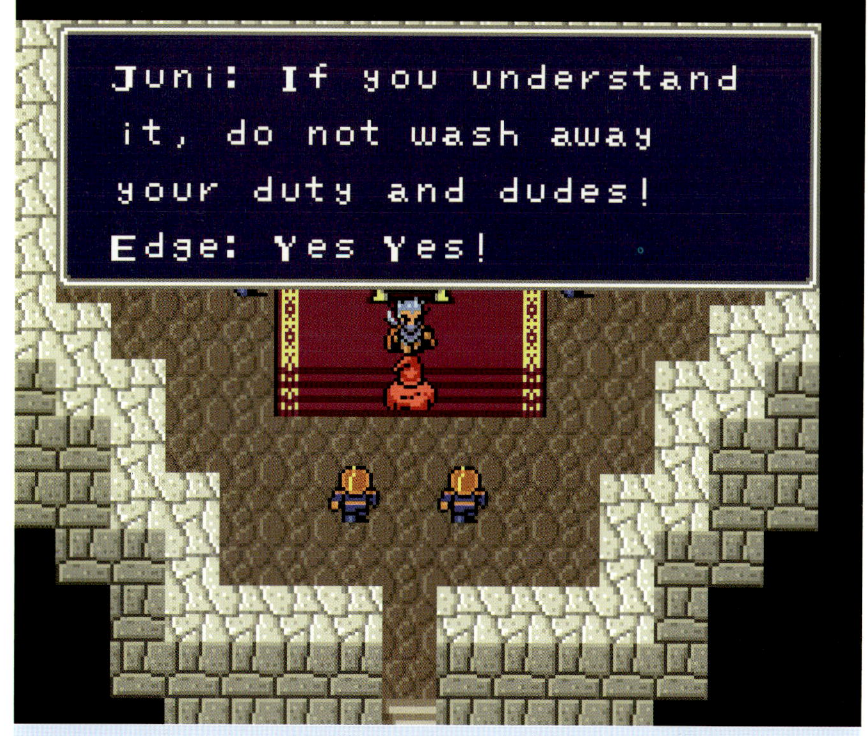

Old Man: If you're aware of that, then you should stop chasing after girls all the time!
Edge: Okay, I get it!

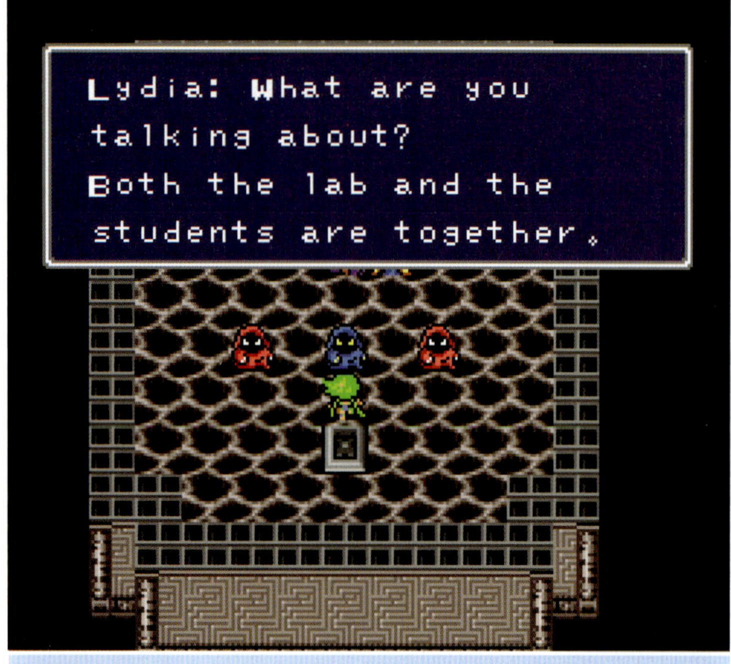

Rydia: Don't be silly. Humans and Eidolons are no different.

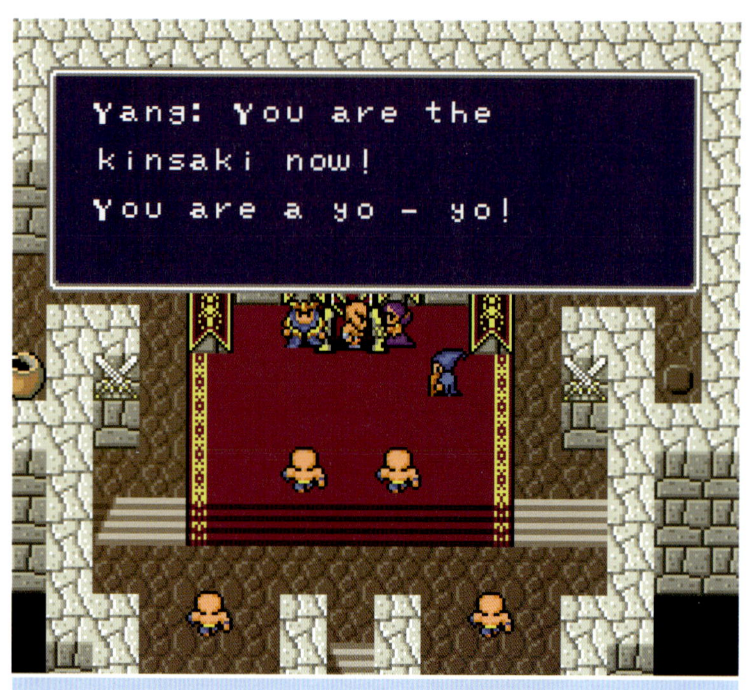

Yang: You're the queen now! Stop calling me "Honey"...!

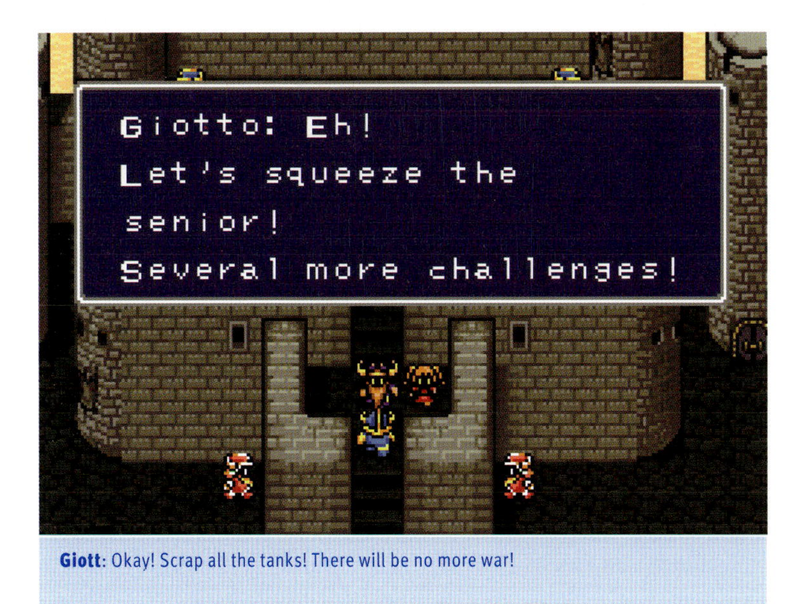

Giott: Okay! Scrap all the tanks! There will be no more war!

Giott: Listen up! Destroy the tanks!

▲ Denture Warriors

In this mistranslation, the Japanese word *sensha* ("tank") was replaced with the word *senpai* ("a person with senior rank") before it was translated. This exact same mistake happened earlier in the game too, so at first I assumed this was a case of a consistent mistranslation across the entire Translation #2 script. Thirty seconds after this line, though, the king's use of the word *sensha* is translated as "squash".

Cid: Really?! In that case, makeup is an important part of being a bride, Rosa!

This is bad! The moon, it's...!

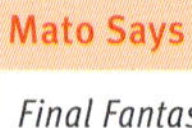

Mato Says

Final Fantasy IV had a lot going for it – a strong story, relatable characters, and a satisfying conclusion. It made players like me realize, "Wait, games can be like this?!"

Final Fantasy IV is unrefined by today's standards, but there were few games like it at the time. Perhaps that's partially why the modern sequels feel so clunky and primitive.

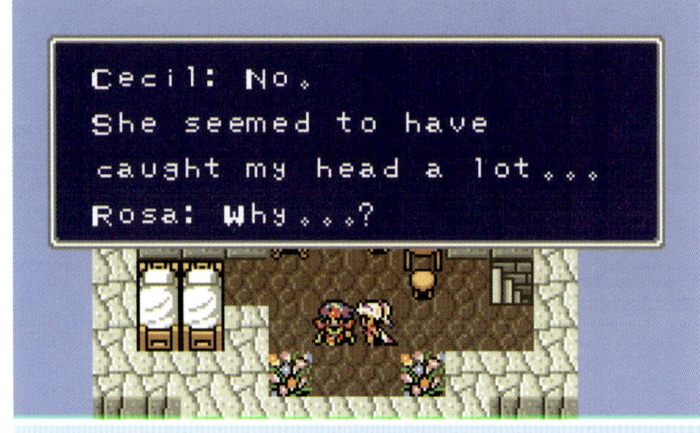

Cecil: It's nothing. I thought I heard my brother's voice...
Rosa: What did he say?

◂ Brotherly Love

Pronouns can be safely dropped from Japanese sentences if it's already clear who's being referred to. When translating such lines into English, the unstated pronouns usually have to be added back into the text. To do this, translators need to be aware of the context and have the ability to read between the lines for relevant clues.

Current machine translations struggle to recognize context at all, which can lead to all sorts of pronoun-related problems. In this example, Cecil clearly refers to his brother in the Japanese text, but the translation's use of "she" immediately afterward indicates that the machine overlooked that information entirely.

◀ Due Credit

The Japanese credits at the end of *Final Fantasy IV* are already in English, so there was no need to translate them for this project. Still, to mark the occasion, I updated the credits to acknowledge the game's new translator. I also included the date I finished playing all the way through Translation #2 for future reference and to give the project a sort of "time capsule" vibe.

Here's hoping this wacky experiment proves useful to future generations of linguists!

So, this thing known as "anger" can make humans stronger...

Final Fantasy IV contains thousands of lines of script text and hundreds of names, so it's unfeasible to cover everything in a single book. Still, I thought it'd be nice to include some "honorable mentions" from Translation #2 that didn't make the main cut. Please enjoy these examples of modern machine translation!

Story Dialogue

Machine Translation	Intended Translation
It is fearful that the stuff of Cecil is getting wet.	Unfortunately, it appears that wretched Cecil harbors suspicions about you.
Cecil: Hell!	Cecil: Your Majesty!
What kind of hell are you going to do? We are sticking to the fish everyone!	What exactly are you planning, Your Majesty? Everyone has grown suspicious of you!
My daughter was also occasionally busy. Although I will end up paying taxes if I study Konku, this one will match me.	My father was a Dragon Knight too. I'm sure I'd rise up in the ranks if I mastered the Dark Sword arts instead, but this suits me better.
There are lots of rice.	Inventory is full
I'm getting clothed tightly!	Take care and fight good, ya hear?!
Porom: What is the meaning of meteor? Palom: Grandpa!	Porom: If he knows about the Meteor spell, then... Palom: Hey, old man!
Was it a fist?	Oh, you're alive.
The magma soup is amazing!	The Magma Stone is glowing red!

Rosa: Cain!?
Cecil: He's back to normal...

Porom: Eheheh! We will be your guide to Mt. Ordeals.

Machine Translation	Intended Translation
Let's hide my wedding ceremony for you.	I will convert my consciousness into the power of light and entrust it to you.
I can not do it when everyone is knocking on me ...	I'm pathetic. While you guys are out there fighting, I'm...
I will congratulate you on the racket!	It's a pleasure to meet you!
It is the Barbarians of the Evangelical!	It's Barbariccia, one of the Four Heavenly Kings!
It is totally painful to be caught up from the sky with a rainbow trouble ...	Our mighty tank squad has trouble fending off attacks from the air...
It was just bruised because of the fact that it was muddy and a muffler earlier.	I'd like to, but it got a bit messed up after that bombardment and crash landing.
Congratulations to your taste!	Maybe this will help you remember!
I was kill ..	I don't expect you to ever forgive me...
But this place is a cousin!	But we've already reported our location!
If you become worried and head for the Damshan, you will be touched by the strange things in this place.	I grew worried, so I'm on my way to Damcyan now, but I'm having difficulty dealing with a giant monster in an underground lake up ahead.
Rosa: It was a boyfriend! Cain: Put your parenthesis on that much	Rosa: Oh, you're all right! Cain: Even after that flashy show of yours...
It's kind of wonderful enough to hold an earthquake ...	Your strength is indeed worthy of self-confidence...
There will be enterprises! This is a bad thing for you!	We have the Enterprise! We can go anywhere in the world in the blink of an eye!
I will return back to Baron after a while on an astronaut.	It's back in Baron, thanks to the power of remote control!
I love you soon.	We met him on Mt. Ordeals.
Do not get caught in a cupboard!	We can't jump in! It's flowing too fast!
Do you have a rice cake?	Do you have a membership card, sir?

Machine Translation	Intended Translation
The monk wanted a lot. All that is still in the morning is still fresh.	Our main monk squad was wiped out. The monks at the castle are not yet fully trained.
When saying how old puppet father works ...	The king has been acting so unusual lately...
Hello I am a lucky charm.	Hello. I'm a traveling scholar.
Everyone has a bad bug at the beginning	Everyone is born as an innocent newborn at the start, and yet...
Would you like it to be something for a cow.	I can have them gather any materials you need to repair it.
It seems that you have come from the otaku of Kintonoyama's plant, but... Now you are occupied and you have to leave it in the sky.	Our ancestors came here from a big cave on top of the northern mountain, but it's sealed up now. Besides, you'd need to fly to get there anyway.
When we return the rice tomorrow, the rice paddies do not get wet	When we return the sun's stone to its home, the path to our homeland shall open....
Oh my God's Grief!	I-it's the ghost of our boss!
Many things are said about the fence, but if you enter into the fence of the injury ...	Everyone is saying things about His Majesty. If he were to hear any of it, though...
I hope that I will not use this deaf again anymore.	I hope I never have to use this prison ever again.
It seems that you will get a ridicule but what you can find is what you are supposed to do.	I hear you're setting sail. Mysterious shipwrecks have been happening a lot lately, so please be careful!
He heard that he received a police box at Hobbs!	I hear you were ambushed on Mt. Hobs!
I'm getting tired of getting down with bad things!	Hurry and beat up the bad guys with your special kick!
I caught scorching the bad guy firmly!	Go give those bad guys a serious beating!
Hello there is something frightening!	If someone with an evil heart were to possess it, the results would be horrifying!
I just hope to hesitate to tear you apart now.	We can only pray now that His Majesty will rest in peace.
I'm toxicologists!	We'll avenge you!

P-p-please, just leave me alone!

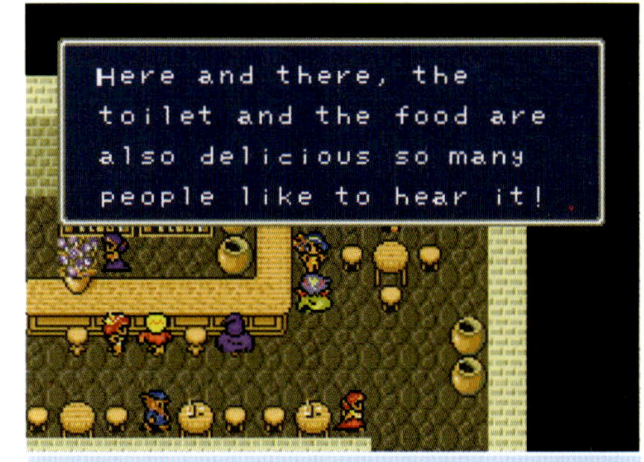

The air and the food here are great, plus it's nice and relaxing and full of women!

Machine Translation	Intended Translation
Please show me as much as a venerable wagon!	Unleash your awesome ninja skills on them and teach them a lesson, Young Master!
If it is a girl you will definitely come and go!	I'm certain the young master will slay our enemies!
I am the most probably okay!	I'm the most ravishing girl in the village!
If there is no key to withdrawing a tournament, I can not enter that animal.	There's a key that can break the seal on the cave. It's the only way you can get inside.
You guys are going to deal with what kind of husband!	In the meantime, hurry to the Sealed Cave southwest of here!
Oh dear? Being squashed!	C-Captain? You look so magnificent!
I've been touched by somebody for leaving me a chick.	He left us in charge of repairing the airships so he could have time to think some stuff over.
It is crocking now!	It's too dangerous to go outside right now!

Dwarf weapons are heavy!

Machine Translation	Intended Translation
Here is the fondness of the finger!	This is His Majesty's bedchamber!
The main thing that everyone is tapping is baked me. Please tell me if there are kimono.	I drew all the monsters that you've been fighting. Let me know if there are any monsters you're especially fond of.
You can make a magazine by miso!	Press right to defend!
What is inside of you, what is it!	My husband did? That's nonsense!
Are you sorry for lots of fun guys?	Are there lots of bad guys on the surface?
Sea cucumber ...	Leave it to me.
Shinjin peas and Toshio: Yawning Yawei. There is a strong yaw. There is no missing yaw.	Toshio Endo, the New Guy: My shoulders are so stiff! My eyes hurt! My ears hurt!
All of these daughter's goods are left undone!	The fate of all life on this world is in your hands!
What I once felt was only when the fish caught in the bell.	His only defeat came long ago when lightning struck his sword.

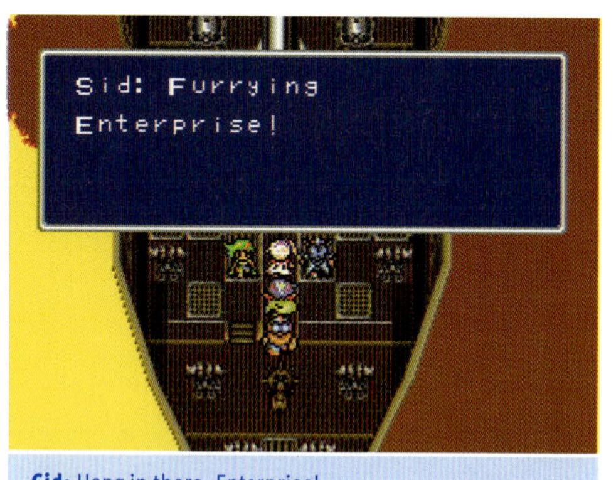

Cid: Hang in there, Enterprise!

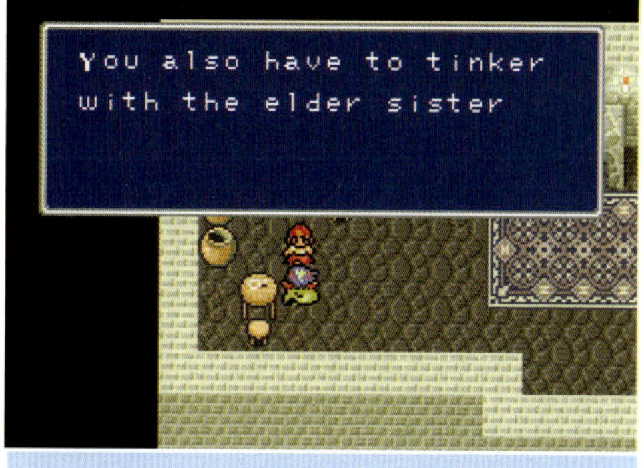

He's fiddling with those airships again, I assume?

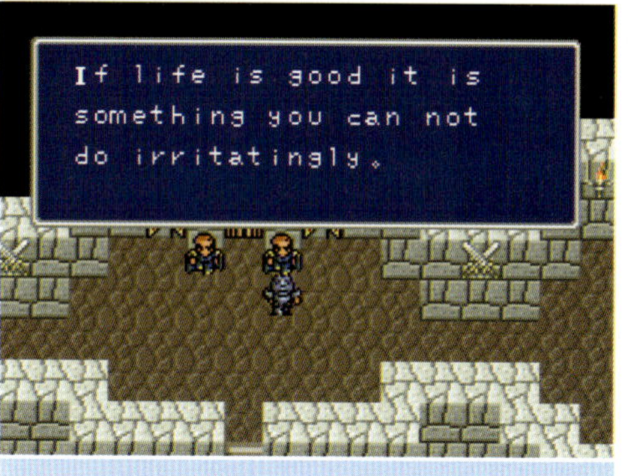

If you value your life, then I suggest you don't try anything funny.

Your men treat 'em too rough.

Machine Translation	Intended Translation
We are mostly in trouble with Zemus from our house.	It is taking all of our energy just to neutralize Zemus' telepathy coming from the inside.
If it is, I also have to replace this girl with that semester!	In that case, I must make this Zemus fellow pay!
However, as long as there is a Jerky mind, I will be sorry.	But evil will never fully disappear. Both good and evil exist in all things.
Welcome. My girlfriend's sideness is bad.	Welcome. Oh, that young girl looks pale.
It is your struggle to shave before that.	Your job is to dispose of him before that happens.
This is a battle against my eccentricity!	This battle is between me and myself!
November!	The Magic Ship!
I want a desire of my son quickly.	I can't wait until I get a desk of my own.
I am about to scrub around in the dusk.	Now it sleeps in the mouth of a dragon.
It will only be what can repel the throbbing of the labor force.	Only he who can reflect the Eidolon God's power can defeat it.

Maybe that old airship guy could modify it?

Battle Dialogue & Messages

Machine Translation	Intended Translation
Opportunity to endure when you express yourself!	Your chance to defeat it is when it takes its physical form!
Mug "Rug now! I am going magical!	Magu: Now, Ragu! Cast a spell on me!
I'm noisy!	Shut up! Be quiet!
The time has come for us to use the meteorology ...	So, the time to cast Meteor has come...
He brought him back to court!	The king and queen regained control of their minds!

(Unshortened Text: "You should be able to squeeze gas!")
Breathe this gas and become living corpses!

Battle Actions & Spells

Machine Translation	Intended Translation
King's cheek	Hellfire
Breath of dress	Mist Breath
Case study	Stone Glare
Job launch	Binding Cold
Medical institution	Restore
Job abstract	Unbind
Sweet potato	Tornado
Curtain	Flamethrower
Loudspeaker	Water Bomb
The lion	Thunder Blade
Bunker	Split Image
Bag	Cover
Lump	Airship
To despair	Steal
Boy	Defend
I do not care	Don't Cover

Throw

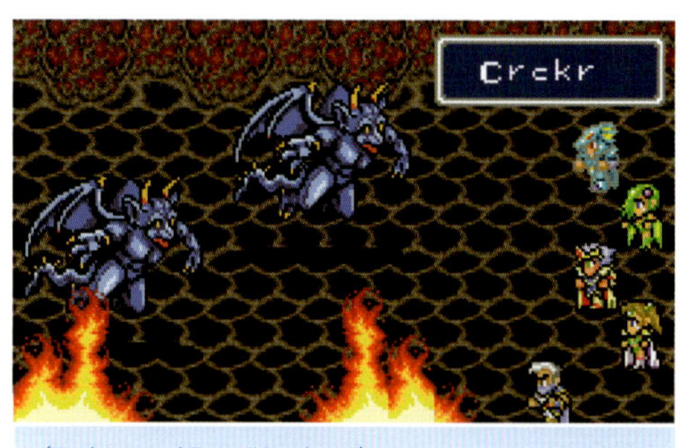

(Unshortened Text: "Crockery")
Fire Bomb

Character Classes

Machine Translation	Intended Translation
English corkscrew	Dark Knight
Cooking	Dragon Knight
White Shrimp	White Mage
Scholarship	Summoner
Testimony	Sage
Apostles	Royalty
A boy	Ninja
Tatami	Lunarian
Cheep	Engineer

Status Conditions

Machine Translation	Intended Translation
Sushi	Stone
Lid	Pig
Tomorrow	Poison
Fault	Confusion
Teacher	Berserk
Shikoku	Death Sentence

Enemies

Machine Translation	Intended Translation
Hand Leg	Hundred Legs
Rudolf Lamb	Baby Rudran
Attendance	Machine Gun
People	Machine Soldier
Attached eyelids	Moon Goddess
Seoul	Soul
Good luck	Iron Knight
Life Paper	Thunder Fish
Ginger roe	Bard

Kiyoshi Yoshī

Stats & Shops

Machine Translation	Intended Translation
Calamity	Strength
A lot	Stamina
Boy	Defense
Magical Wonderland	Magic Evasion Rate
Crow	Buy
Vegetarian fried rice goosecho I do not care as long as I do not care	Your inventory is a mess, sir. You need to organize your items before you can take this.
There is no money for customers. I feel troubled.	You don't have enough money, sir. Please stop fooling around.

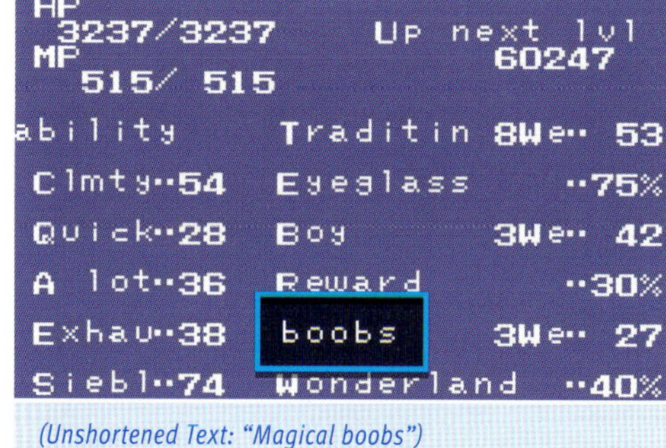

(Unshortened Text: "Magical boobs")
Magic Defense

Items & Equipment

Machine Translation	Intended Translation
The claw of a poster	Lightning Claw
Nail clippers	Fairy Claw
jealous claw paw	Hell Claw
Hot rod	Fairy Rod
Chicken stick	Power Staff
What's next	Wave Staff
The King of Earthquakes	Sage Staff
Tonight's sword	Legendary Sword
The Sword of Affection	Ancient Sword
It is not it	Kunai
Children	Kitchen Knife
Clowns	Wooden Hammer
Hammerhead	Shock Whip
Women's samurai	Fuma Shuriken
english prince	Chakram
Powered by Translate	Blank
Good Awake	Yoichi Bow

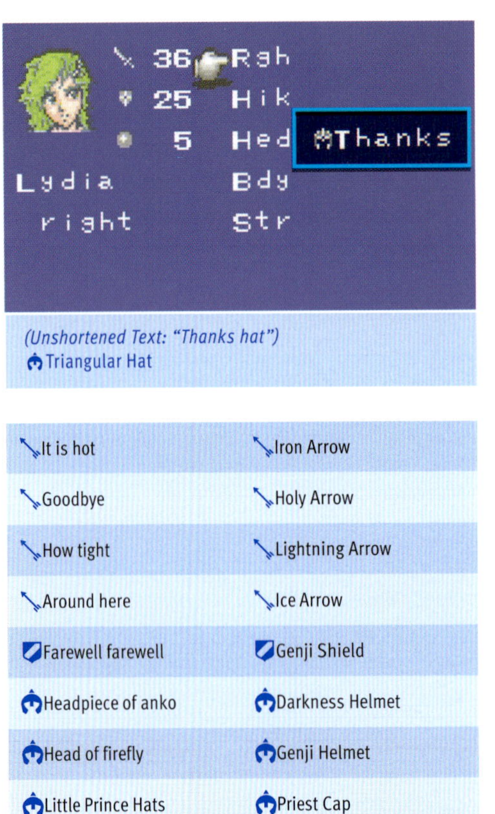

(Unshortened Text: "Thanks hat")
Triangular Hat

It is hot	Iron Arrow
Goodbye	Holy Arrow
How tight	Lightning Arrow
Around here	Ice Arrow
Farewell farewell	Genji Shield
Headpiece of anko	Darkness Helmet
Head of firefly	Genji Helmet
Little Prince Hats	Priest Cap
Kimchi of Kin	Gold Hair Ornament
Torsion chirping	Headband

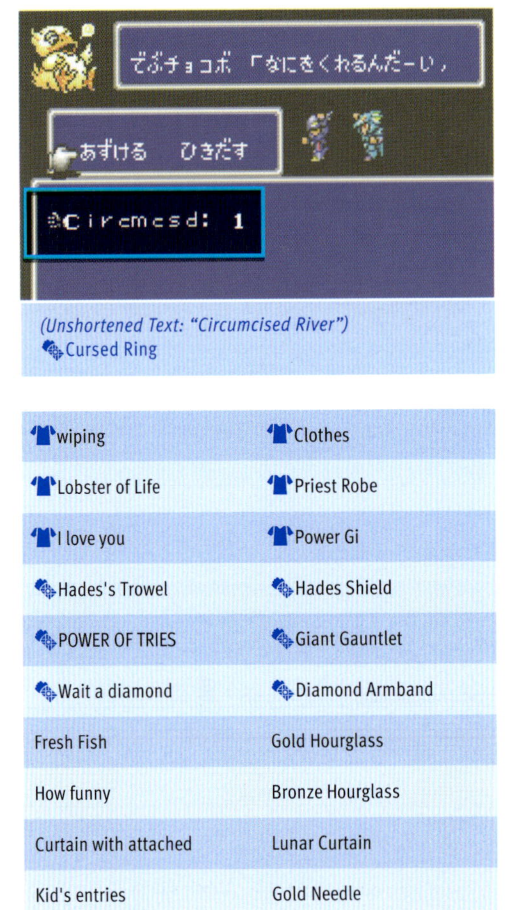

(Unshortened Text: "Circumcised River")
Cursed Ring

wiping	Clothes
Lobster of Life	Priest Robe
I love you	Power Gi
Hades's Trowel	Hades Shield
POWER OF TRIES	Giant Gauntlet
Wait a diamond	Diamond Armband
Fresh Fish	Gold Hourglass
How funny	Bronze Hourglass
Curtain with attached	Lunar Curtain
Kid's entries	Gold Needle

Locations

Machine Translation	Intended Translation
3 time	3rd Floor
4 yes	4th Floor
6 days ago	6th Floor
8 years old	8th Floor
Bullet	Weapon Shop
Let's face it	Underground Passage
Chocobo starch	Chocobo Forest
Chocobo's unevenness	Chocobo Village
Husband and wife	Hidden Room
Tatami noodle shit	Cave of the Lunarians
I want to talk to you	Underground Waterfall
Damian's Deaf and	Damcyan Prison
Anthrion's fish	Antlion Hive
The hobby of the hobbs	Mt. Hobs Entrance
Countercurrent clothes	Magnetic Cave
Babuil's Eggplant	Tower of Babil 1F
Eye of the cricket	Giant's Mouth

(Unshortened Text: "Birth of a baby B3")
Lunar Underground Ravine B3

Machine Translation	Intended Translation
Not at all	Magic Ship
Dwarves' mortality	Dwarf Base
Several variations of the Zen shinkin	Eidolon God's Cave
Accompanied by a baby	Lunar Underground Ravine
Chopsticks Babuil	Underground Babil
Cheerfulness	Nap Room
Language instruction manual	Eidolon Library
Lydia's squirrel	Rydia's Home

(Unshortened Text: "Shibuma Police box")
White Magic Laboratory

(Unshortened Text: "Kurouma People's Court")
Black Magic Laboratory

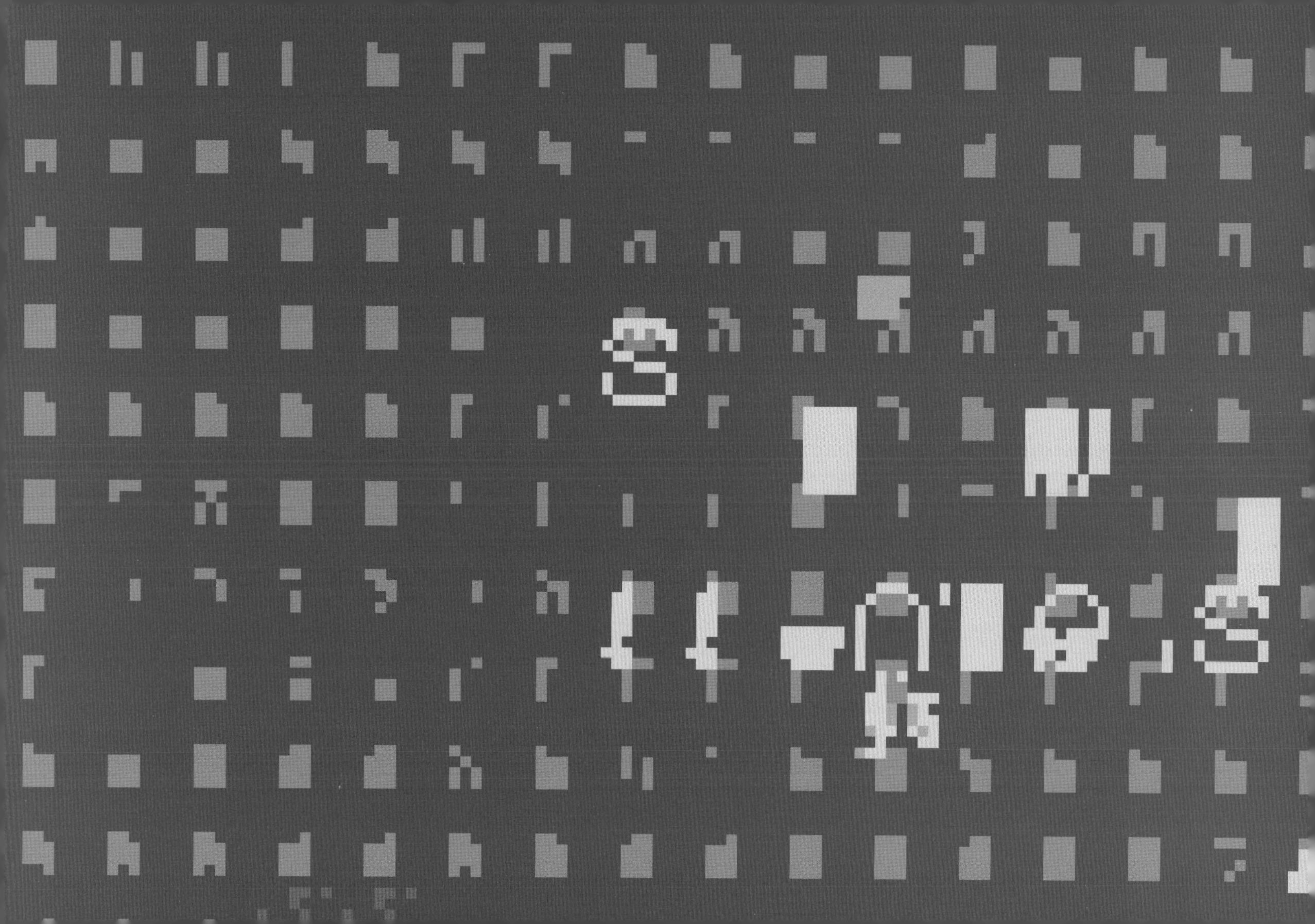

Translation #1 vs. Translation #2

Over the past 180 pages, we've looked at two types of machine translation and how they handled a large-scale Japanese-to-English game translation. While both approaches made many of the same missteps, they also produced very different translations. So let's take a moment to compare Translation #1 and Translation #2 and their patterns.

Translation #1's characteristics:

- English grammar is almost always complete nonsense
- Context is completely ignored
- There's no understanding of what's being translated as a whole
- The purpose of the translation as a whole is completely ignored
- Short bits of text are sometimes left untranslated but written out in English
- Despite poor translation quality, the original intent sometimes remains intact

Translation #1 Patterns

The machine responsible for Translation #1 used an old-school phrase-parsing approach to translation – in other words, it would break sentences apart into single words or groups of words and try to translate them one by one. On the surface, this probably sounds like a logical idea. But, as we've seen, it's actually a primitive technique that leaves a lot to be desired.

In the end, there's one question that matters most: could a newcomer play this translation and understand – in some small way – how to progress through the game? Although Translation #1 is a miserable pile of nonsense, there's just enough in the main text for players to grasp what's going on and where to go. Secondary text is handled just well enough that players can navigate menus without resorting to trial and error. With all of this in mind, I feel Translation #1 sits on the "minimally, barely playable" level.

Translation #2's characteristics:

- English grammar is surprisingly good
- Context is almost always ignored
- Japanese words are often replaced with similar-sounding but very different words before translation
- Some correctly translated words are replaced with unrelated English words that start with the same letter
- Some Japanese words aren't translated at all; instead, they're replaced with random English words that start with the same consonant
- Unique names are rarely translated the same way twice
- Entirely new English words are invented out of thin air
- Large groups of text are sometimes left untranslated but written out in English
- Some Japanese words are left untranslated and left in Japanese writing
- Phrases sometimes "echo" and repeat themselves mid-sentence for no reason
- Despite the smooth English grammar, the meaning of the original text is usually 100% destroyed

Translation #2 Patterns

The latest trend in machine translation relies on a type of artificial intelligence that can grow and learn on its own. Instead of looking at words as isolated blocks of text, the neural network behind Translation #2 seemed more interested in the *ideas* behind words.

In practice, this alternate approach to translation should produce more meaningful results – in fact, it's how most professional translations are handled. Unfortunately, this particular A.I. struggled to recognize any ideas at all. As a result, even though Translation #2 reads more smoothly than Translation #1, it suffers from many more problems.

Again, all of this boils down to one important question: could a newcomer play this translation and understand – in at least some small way – how to progress through the game? In my opinion, the answer is clearly "no". Clues, names, and story events are so nonsensical in Translation #2 that newcomers would have to rely entirely on guesswork and trial-and-error to play through it. As unintentionally entertaining as it is, Translation #2 has so many issues that it's hard to even call it a translation at all. In many cases, a random sentence generator would be just as accurate.

If you paid someone thousands of dollars to translate a game, would you accept Translation #1? What about Translation #2? If you had to, which would you be more willing to pay for?

Taking Translation #2 Even Further

As we've seen, the Japanese language has three separate writing systems: *hiragana*, *katakana*, and *kanji*. Normal Japanese writing uses a balance of all three systems, but many old games like *Final Fantasy IV* lack *kanji* for technological reasons. Although *kanji* isn't needed to write in Japanese, it helps in three big ways:

- *Kanji* avoids potential confusion
- *Kanji* marks where words start/end
- *Kanji* says a lot with few characters

Basically, Japanese sentences are riddled with tricky, confusing traps. Context can guide you around some of them, and *kanji* can clear out most traps entirely.

So here we have a Japanese game without *kanji*, being translated by a machine that can't understand context – of course it's destined for disaster. But what if we rewrote the script to include *kanji*? How much better would the machine translate it then?

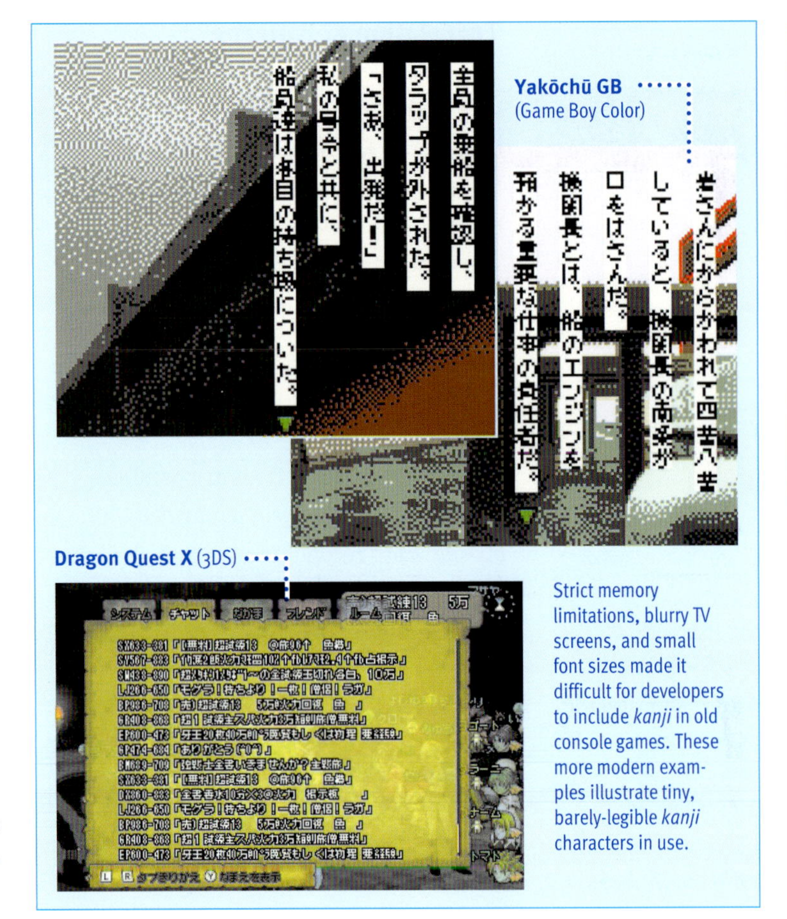

Yakōchū GB (Game Boy Color)

Dragon Quest X (3DS)

Strict memory limitations, blurry TV screens, and small font sizes made it difficult for developers to include *kanji* in old console games. These more modern examples illustrate tiny, barely-legible *kanji* characters in use.

Mato Says

In overly simplistic terms, the balance of *hiragana*, *katakana*, and *kanji* in everyday Japanese is like the balance between uppercase letters, lowercase letters, and spaces in English. English text gets a tad harder to read if you don't use lowercase letters, for example, and it gets exhaustingly difficult if you don't use spaces. But using all three makes for easy reading.

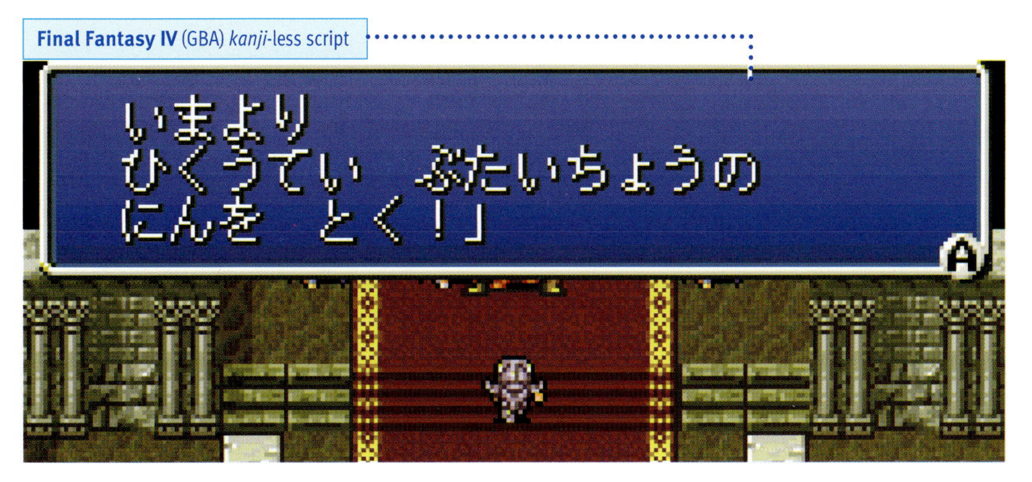

Final Fantasy IV (GBA) *kanji*-less script

Final Fantasy IV (GBA) *kanji* script

As luck would have it, some Japanese re-releases of *Final Fantasy IV* include an option to switch between two scripts: the original *kanji*-less script and a revised script that adds *kanji*.

Out of curiosity, I ran the *kanji* script through Translation #2's machine translator to see how it performs. Its translations are indeed better, but many of the same illogical patterns still appear with regularity.

In simpler terms, both Translations #1 and #2 would've turned out better if I'd used the *kanji* script found in later versions of *Final Fantasy IV*. I knew going into the project that the original script would lead to an inferior translation, but I also wanted to see if modern machine translation systems could handle a script that Japanese five-year-olds can read. If this were a race, the five-year-olds would be winning and the machines would accidentally be running the wrong way.

It's no exaggeration to say that Japanese first-graders can read better than cutting-edge neural network machine translators. In this example, we have an all-*hiragana* paragraph taken from an actual first-grade textbook. Despite the simplicity of the text and its helpful formatting, the machine translator fails to read any of it properly. As a result, a story about seashells and a cute little bear is replaced with complete nonsense about fences, canes, and bombs.

ももいろの　かいがらは、にばんめに
きに　いって　いた　ものなのです。
くまの　こは、どう　しようかと　おもいました。
そして、かいがらを　そっと　しまって、
うちへ　かえりました。
その　よる、くまの　こは、
いっしょうけんめい　かんがえました。
つぎの　ひ、くまの　こは、

"The Momoiro 's fence was the one going to Bomb. I thought what to do for the bear. And gently stashed the cane and went back home. Because of this, the bear crisped together."

An excerpt from *Atarashī Kokugo 1*
Machine translated by Google Translate

This random excerpt from a popular children's book also proves problematic for state-of-the-art technology:

"Alm's life is just about new things. There are things that are savory. By the time the paddy ended, Heidi was getting squeezed as well. I can do things like cheese making. Heidi is already a mountain breast."

An excerpt from *Heidi, Girl of the Alps*
Machine translated by Google Translate

アルムでの　くらしは、あたらしい
ことばかり。あらしに　あったり、
ときには、きけんな　ことも　あります。
夏が　おわるころには、ハイジは、
やぎの　ちちも　しぼれるように
なって　いました。チーズづくりの
おてつだいだって　できます。ハイジは
もう、すっかり　山の子です。

(24)

Comparisons

Now that we've examined Translation #1 and Translation #2, let's see how they compare side-by-side under different circumstances. We'll also see how Translation #2 would've looked if the original Japanese script had included *kanji*. For simplicity's sake, we'll call this alternate version "Translation #2.5".

Simple Sentences

Translation #1 has phrasing issues with this basic Japanese sentence, but the key words remain intact. Translation #2 is wrong in every way. We see from Translation #2.5 that *kanji* improves the situation considerably – in fact, it results in a 100% valid translation of the line. The machine simply had no way of knowing that "Bomb" was the intended word and not "Bom".

In this instance, Translation #2.5 is the best of the bunch, followed by Translation #1. Translation #2 isn't even a translation.

Intended Meaning

Translation #1

Translation #2

Translation #2.5

Tricky Sentences

The original Japanese version of this line uses simple grammar that's only marginally more complex than the previous example. But it also features a few tricky words that are easy to misread.

The English grammar in Translation #1 is very poor, but thanks to the words "arrogance", "elder", and "do not", the key concepts remain fairly intact.

Translation #2 is absolute insanity. And although Translation #2.5 wipes away that ridiculousness, we can see that adding *kanji* to the Japanese script doesn't help much in conveying the original meaning.

In this instance, Translation #1 is the clear winner, with Translation #2.5 coming in at a distant second. Again, Translation #2 isn't even a translation.

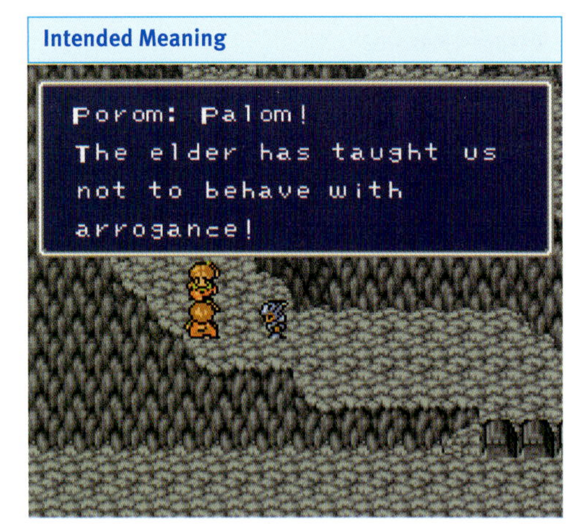

Intended Meaning

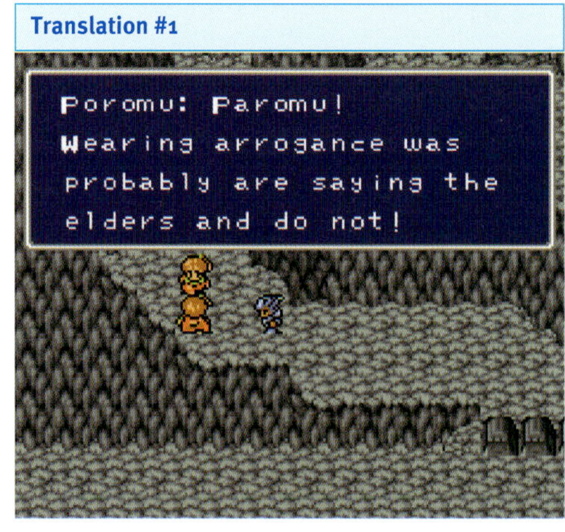

Translation #1

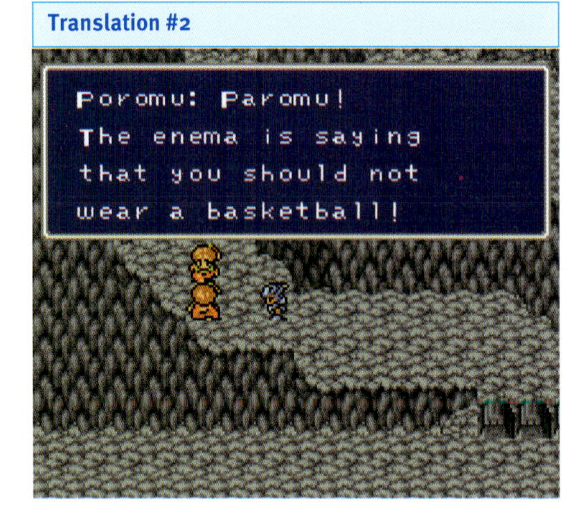

Translation #2

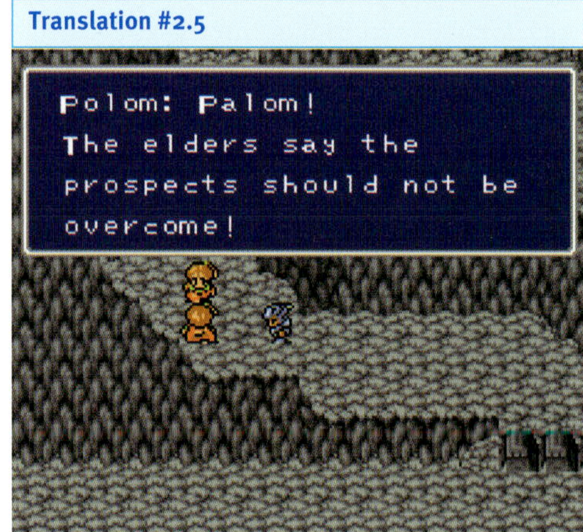

Translation #2.5

Idioms

The Japanese version of this line looks simple on the surface, but it actually contains an interesting idiom: 油を売る (*abura o uru*). Literally, this Japanese phrase means "to sell oil", but it functionally means "to slack off" or "to waste time chatting about nothing important".

In Translation #1, we see that "oil" was translated as "fat", which is a perfectly proper translation in discussions about dietary fat or animal fat. Elsewhere in the sentence, a confusing homophone caused the machine to translate "sell" as "hit".

Translation #2 makes little sense in any way. Similarly to Translation #1, the word for "sell" was misread as the word for "hit", which led to the word "blow" being used. To make things worse, the sexually suggestive phrase "blow me" takes the translation even further from the original meaning.

Intended Meaning

Adding *kanji* to the Japanese script makes Translation #2.5 shine in comparison – it includes a perfect, literal translation of the Japanese idiom. Unfortunately, the literal translation still sounds nonsensical if you don't know Japanese already.

In this instance, all three translations fail. Translation #2.5 gets close, but can't make the jump from literal meaning to functional meaning.

Translation #1

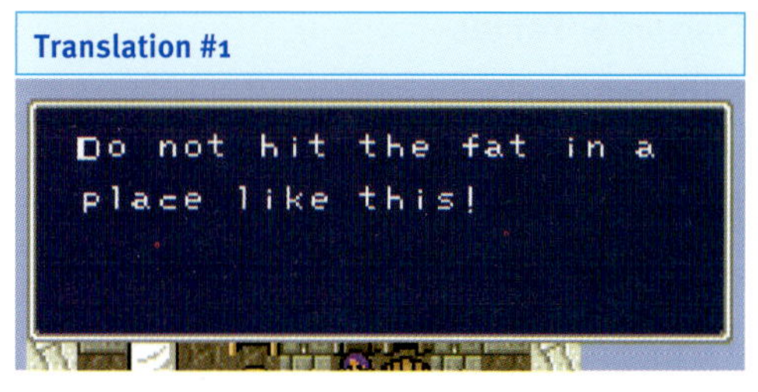

Translation #2

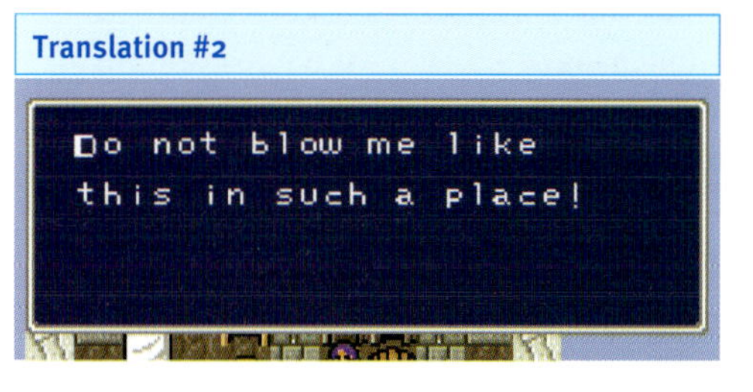

Translation #2.5

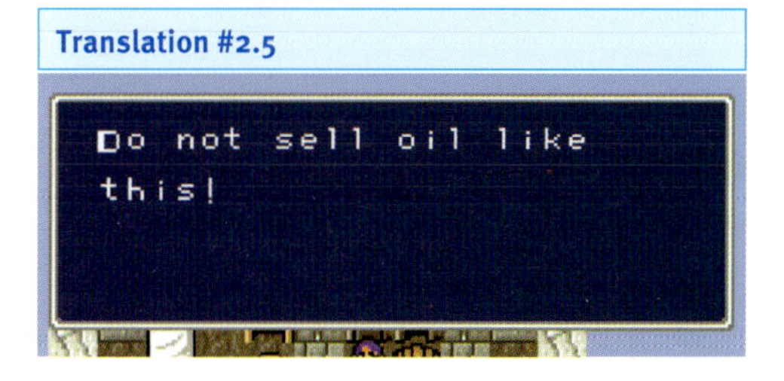

Names & Terms

In this example, we have ten item names in a menu. As we've seen, names can be tough to translate without context, and issues like name length limitations can make the process even tougher.

In Translation #1, most of the item names make sense. Two names remain in Japanese, and one is mistranslated altogether: "Pesticide".

In Translation #2, every name is mistranslated or left in Japanese.

In Translation #2.5, no names are left in Japanese, but two are mistranslated. Even with *kanji* to help, Translation #2.5 follows the same patterns we saw throughout Translation #2, resulting in an item called "Launch Grommets".

Overall, Translation #2 continues to challenge the minimum definition of what a translation is. Meanwhile, thanks to *kanji*, Translation #2.5 manages to reach Translation #1's level.

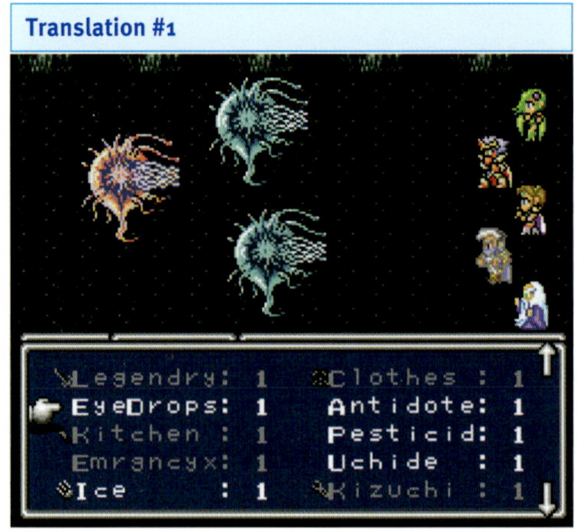

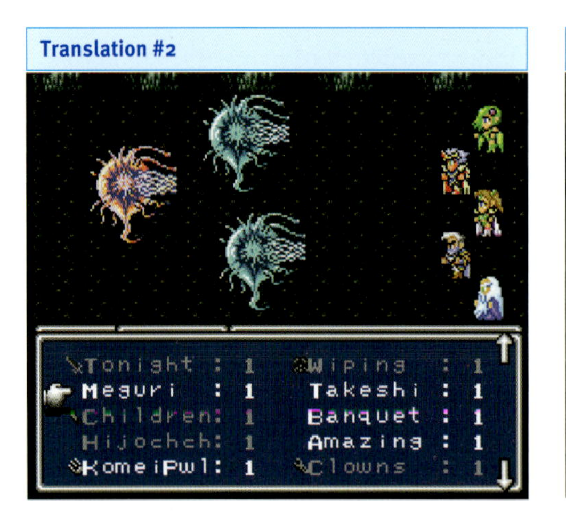

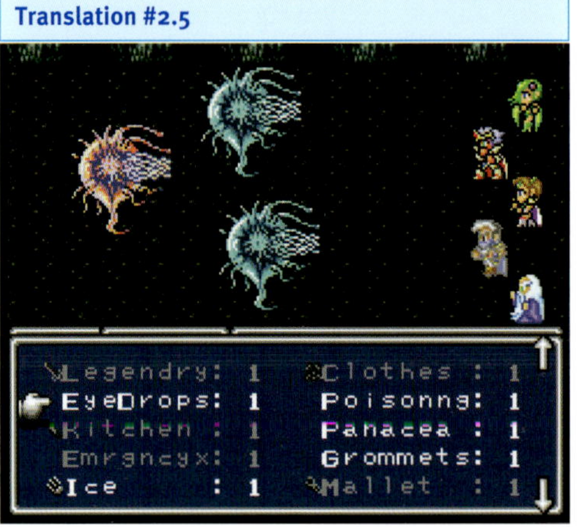

Summary

After looking at Translation #1 and Translation #2 in detail, it's clear that both translations are far from acceptable for gamers and publishers alike. However, it's also clear that Translation #2 is a significant step back from Translation #1, despite its use of cutting-edge neural network technology.

With some experimentation, we also learned that Translation #2 can perform better when the source text is revised and reformatted in a certain way. At the same time, this indicates a reading comprehension level beneath that of Japanese first-graders.

With all of that said, however, this was an isolated experiment that doesn't represent how well neural networks translate everything. In this instance, we've seen how well a neural network can translate a full-scale Japanese video game, but it's likely that the same neural network can adequately translate Spanish newspaper articles into English or French e-mails into German, for example. What's more, neural networks are designed to grow and improve over time, so it's possible that the translation problems we've seen in this *Funky Fantasy IV* project will eventually be resolved.

Machine translations have long had a reputation for being unreliable, and there have been enough horror stories over the years that savvy publishers understand the risk of translating games with automated tools. Of course, such tools do have their place, and sometimes automated translation is the only feasible option. At the very least, though, I hope this little project has helped explain what can go wrong during machine translation and why.

On a side note, a research paper about machine translation suggests that Google's neural network system reduces translation errors by an average of 60%. I'm not sure what percentage Japanese-to-English translation improved, but I estimate that Translation #2 actually featured 50-100% *more* translation errors than Translation #1.

> Using human-rated side-by-side comparison as a metric, we show that our GNMT system approaches the accuracy achieved by average bilingual human translators on some of our test sets. In particular, compared to the previous phrase-based production system, this GNMT system delivers roughly a 60% reduction in translation errors on several popular language pairs.

Google's Neural Machine Translation System: Bridging the Gap between Human and Machine Translation (October 8, 2016).

Aftermath

Although my *Funky Fantasy IV* machine translation experiment began as a small part of something much bigger, it quickly took on a life of its own. And, before I could even blink, it had gathered a surprising amount of attention.

The Project Goes Public

During *Funky Fantasy IV*'s development, I occasionally shared images with friends and *Legends of Localization* fans. I also played through the machine translations from start to finish collecting images and information. On a whim, I decided to stream the entire experience online for other gamers' amusement. Afterward, I gathered those screenshots and videos together, made a small web page about the project, and then called it a day.

I had learned a surprising amount from this fun little experiment. Little did I know there were more surprises in store!

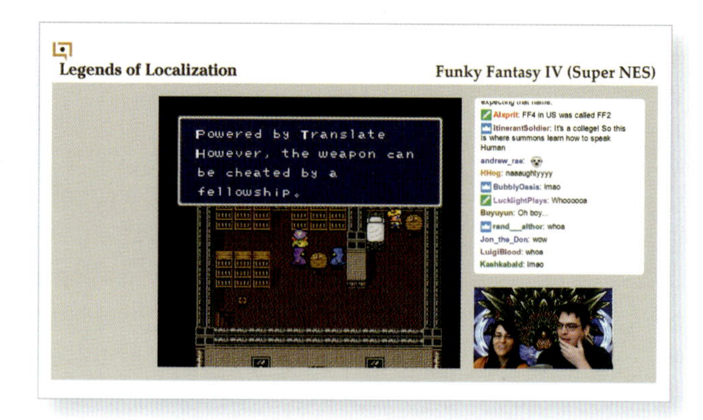

Online Response

The *Funky Fantasy IV* page generated an immediate flurry of interest online. Reporters quickly pounced, hoping for exclusive interviews about the project. Big-name tech sites, game sites, and blogs published articles about *Funky Fantasy IV*. Word of the 100% machine-translated game rocketed across all the social media channels. And although it was never the goal of the project, translation sites around the world cited *Funky Fantasy IV* as an example of why human translators are better than automated tools like Google Translate. Even now, nearly a year later, *Funky Fantasy IV* attracts lots of online attention.

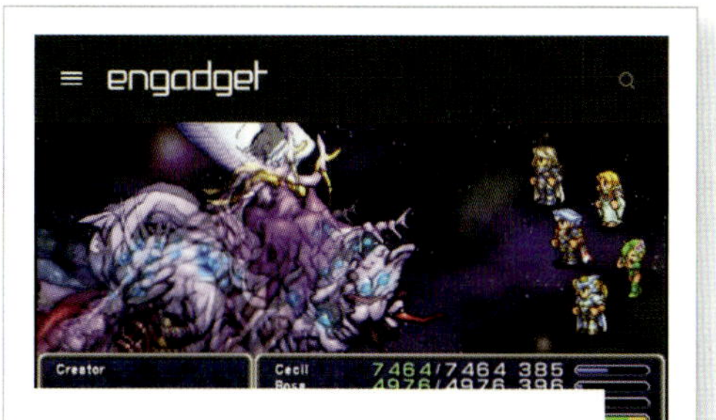

Someone Google Translated 'Final Fantasy'

Prepare yourself for an even more confusing JRPG.

Tom Regan, @grapedosmil
12.13.16 in AV

As much as we love them, Japanese role-playing gam
baffling at the best of times. Yet thanks to some cleve
localization, teams of writers and translators around
have managed to make sense of these intriguing adve
what if these localization teams didn't exist? That's tl
translation enthusiast Clyde Mandelin asked, resultir

I woke up one morning and suddenly there was a deluge of articles about some guy who Google Translated *Final Fantasy IV*. I don't know who that guy was, but I think he was a crazy person.

FINAL FANTASY

What Happens When You Ask Google to Translate 'Final Fantasy'

Patrick Klepek
Dec 13 2016, 1:19pm

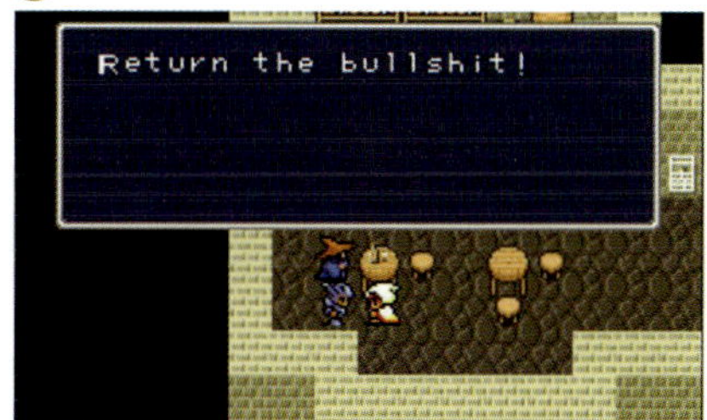

There's a reason humans are still involved in translating video games, but it doesn't mean there aren't lessons to be learned.

Translating a game from one language to another is a nuanced process, involving specialists who carefully parse how words convey meaning, tone, and emotion. It's not as simple as taking the text, dumping it into Google Translate, and seeing what the software spits out. If you did that, as translator Clyde Mandelin has been doing recently with *Final Fantasy IV*, you might end up with a character shouting "The enema is saying that you should not wear a basketball!"

News of the *Funky Fantasy IV* project spread to outlets around the world. I don't know what this article is saying – guess I better run it through Google Translate!

ข้ามกำแพงภาษา! เกมเมอร์เขียนโปรแกรมแปลเกมส์ Final Fantasy IV ด้วยกูเกิ้ลแปลภาษาโดยตรง

เมื่อ : 14 ธันวาคม 2559 ผู้เข้าชม : 1,769 เขียนโดย : Admin VerB

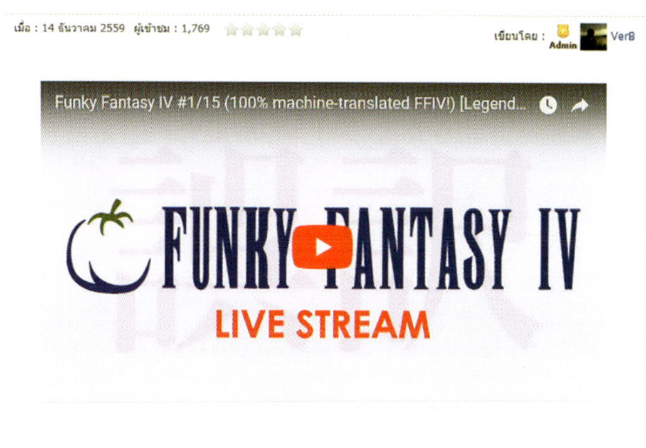

เกมส์ประเภท JRPG เป็นเกมส์อีกหนึ่งประเภท ที่ได้รับความนิยมเป็นอย่างสูงในวงการเกมส์ แต่อุปสรรคของเกมส์ประเภทนี้มีอยู่อย่างเดียว ก็คือ ค่ายเกมส์ญี่ปุ่น มักจะทำเกมส์จำพวกนี้ ออกมาให้เล่นกันในภาษาญี่ปุ่นเท่านั้น จึงทำให้เกมเมอร์หลายๆ คนยอมรับความพ่ายแพ้ ของกำแพงทางด้านภาษากันไป

แต่ก็ยังมีผู้ที่ไม่ยอมแพ้อย่าง **Clyde Mandelin** เกมส์แคสเตอร์บน YouTube ที่ล่าสุดได้ทดลองเขียนโปรแกรมให้ Google Translate สามารถแปลเนื้อหาต่างๆ ในเกมส์ Final Fantasy IV ได้โดยตรง โดยโปรเจคนี้ได้ใช้ชื่อว่า **Funky Fantasy IV** และนี่ก็คือตัวอย่างของโปรเจคที่พวกเขาได้สร้างขึ้นมา

Academia Response

I received lots of e-mails during the initial wave of *Funky Fantasy IV* news articles. One in particular came as a big surprise: I was invited to submit an academic paper about the project!

I've been out of university for many years now, and I've never once considered writing a paper for an academic journal. That stuff is for fancy-pants scholars, not a casual like me.

Still, I realized that this was a once-in-a-lifetime opportunity, so I gave the invitation some thought. Eventually, I decided to go for it, mostly because I loved the idea of the phrase "Funky Fantasy IV" being in an academic journal, not to mention all of the choice quotes from Translation #2. Just think – scholars a hundred years from now could stumble upon a paper about basketball enemas and "Return the bullshit". Now that's the kind of mark I want to make on the world!

Dear Clyde Mandelin,

I just read and liked your post "Funky Fantasy IV: A Machine-Translated Video Game Experiment" on http://legendsoflocalization.com/funky-fantasy-iv/

As editor-in-chief of Germany's oldest and most-influential journal on translation studies I'm always on the look-out for interesting articles for our international readership (we publish in print and online in German and English). Your topic and approach are certainly most interesting and absolutely relevant. Of course you could say that your posts are already as visible as one get get, but within the community of translation studies researchers you might not yet be quite as well-known as you deserve. So, I'm thinking that it would be worthwhile for you reword your post a bit to turn it into an article in the usual style of such publications. Our articles are, of course peer reviewed, but I am quite sure that this will be no problem.

More on our journal here: https://www.degruyter.com/view/j/les (the language can be switched to English at the top right corner)

So, are you interested?

Best regards
Peter A. Schmitt

Prof. Dr. Peter A. Schmitt
Lebende Sprachen – Herausgeber

http://www.degruyter.com/view/j/les

After some free time opened up, I studied other academic papers and built a framework for my own paper. During this, I also learned that academic papers have a lot of bureaucracy involved, and that there are all kinds of fees associated with submitting papers. I looked into it and realized I couldn't sensibly afford those fees, so I wound up shelving my paper. I was only about 5% done with it, so it wasn't a big loss, but maybe I'll give it another shot if the financial roadblock clears up. At the very least, the invitation e-mail and my unfinished paper will always bring back the fondest of memories.

Even though I was only 5% done with my academic paper before I set it aside, the work I put into it turned out to be very useful for this book. In all honesty, I'll probably never pick the paper back up again, but that doesn't mean all hope is lost – maybe modern scholars will cite this book in their own research papers! Seriously, I just really, really want some future scholar to find my research and say "WTF?!" centuries from now.

Industry Response

Weeks after finishing my *Funky Fantasy IV* machine translation project, I was *still* thinking about some of the silliest lines in Translation #2. The project had given me so much entertainment and insight that I decided to reach out to the Google engineers behind the new translation system and thank them.

Google is such a big company that it was hard to figure out who to contact, but the *Legends of Localization* team and I managed to pin down an address. We mailed the engineers a short thank-you letter and included some of our favorite screenshots from Translation #2 for their amusement. We also included several boxes of delicious chocolates as further thanks – and as a way to ensure the letter wouldn't get lost in a pile of random paperwork.

A few weeks later, out of the blue, I received a reply back from a member of Google's neural network project. He sent his own kind thank-you and even offered to chat over lunch if I was ever in the area. I haven't taken him up on the offer, but it was a nice surprise to get a reply at all!

Dear the fine folks behind the Google Translate update,

Hello! As a professional translator and programmer I'm always interested in projects that combine both fields. My most recent project was an experiment I've wanted to do for years: see how well a machine can translate an entire plot-heavy video game.

For this, I chose the all-time classic Japanese RPG, Final Fantasy IV, and turned to Google Translate for the translation work. By sheer luck, I completed the project exactly as the switch to the neural network system was put in place, so I have builds from before and after the switch. The results have been fascinating... and the new translation system created a script beyond my wildest dreams. I've included a few screenshots with this letter, but the entire 20-30 hour game was like this. It was a real learning experience, and a hoot as well!

I'd love to learn more about the different machine translation processes and how the translation choices in this project wound up the way they did. I realize you're all very busy, though, especially for a silly thing like this, so I at least wanted to send this your way for your amusement.

I truly want to thank you for letting this moment in my career happen. I'm not being sarcastic at all, either - this is literally a project that I'll never forget for the rest of my life. Thank you again, and maybe I'll give it another whirl in a year or two and see how things have changed!

Clyde Mandelin
clyde@legendsoflocalization.com

legendsoflocalization.com/funky-fantasy-iv

Subject: your present and letter

Hi Clyde,

I was just alerted that somebody had put two boxes of candy and letters into our microkitchen (not sure how exactly this happened...).

Anyway, thank you very much! It's great to hear from users and if it is a success story like this it makes it even better. If you are in the bay area at any time, we could have lunch at Google once if you are interested, and we can chat more about translation or other things (just send me E-Mail).

I am not sure how much you have seen from our public posts, but here are a few which may be interesting to you as they talk about the translation project (you may have seen them already):

First blog post:
https://research.googleblog.com/2016/09/a-neural-network-for-machine.html

Second blog post:
https://research.googleblog.com/2016/11/zero-shot-translation-with-googles.html

NYT magazine article:
https://www.nytimes.com/2016/12/14/magazine/the-great-ai-awakening.html?_r=0

Again, thanks a lot (everybody loved your letter!)

Best,

Summary

In all, the response to my spur-of-the-moment *Funky Fantasy IV* project was beyond anything I ever imagined. I figured it would maybe get some laughs from some friends, but the overwhelmingly positive response from so many people encouraged me to dig deeper into the project and write a book about it. I hope you've found it as entertaining and educational as I have.

Final Thoughts

We've done a lot during the course of this book. We've taken a detailed look at two large-scale machine translations of a video game. We've witnessed translation technology evolve before our very eyes. And, despite its crummy performance during this project, we've seen how much potential the newest translation technology holds. As a translator and as a programmer, I have absolutely no doubt that machine translation quality is about to soar to a higher level than ever before. I can feel it.

Although this has been a very laid-back look at a complex topic, I hope it's been a fun, informative ride and that it's highlighted some of the challenges that translators – whether human or machine – have to grapple with every day. And, even though this book only offers a single, tiny snapshot of the evolution of translation technology, I would be incredibly flattered if future researchers find it helpful in some small way.

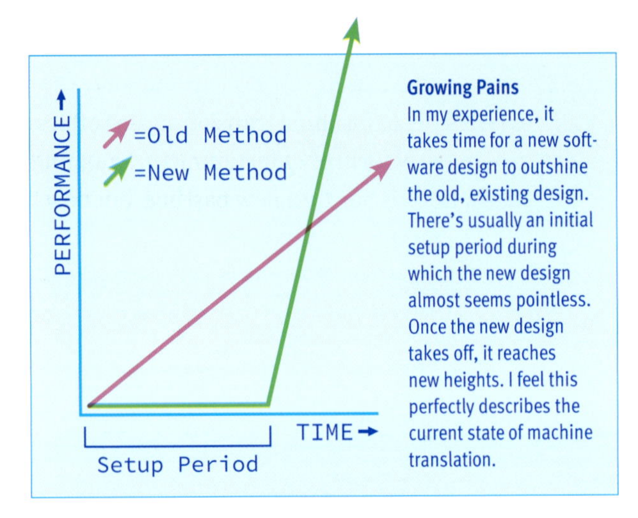

Growing Pains
In my experience, it takes time for a new software design to outshine the old, existing design. There's usually an initial setup period during which the new design almost seems pointless. Once the new design takes off, it reaches new heights. I feel this perfectly describes the current state of machine translation.

For professional translators, the improvement of translation technology opens up many questions. How can we use technology to enhance our own translations? What will the machine translation market look like when competing technologies clash? Will machine translation eventually gain a better reputation? And, of course, the big question: will human translators become obsolete someday? I look forward to seeing what the future holds!

All of that academic and technical discussion aside, I've realized that the current state of machine translation also offers something else: a surprising new form of entertainment. Laughing at bad machine translations is hardly a new pastime, but now that advanced A.I. is involved, the results are quirkier than ever before. As we've seen, modern machine translations are a great way to breathe new life into old games. In fact, why stop at games? Modern machine-translated movies, books, and more can be just as fun!

As this example from *Yoshida-Sensha no Gēmu Manga Taizen* illustrates, machine-translated manga takes on an artistic quality that borders on Surrealism.

Machine translations can bring new laughs to all your old favorite films. In this example, translating the Japanese subtitled version of *Ghostbusters* back into English changed "Let's show this prehistoric bitch how we do things downtown." into something you'd hear in a sex education film.

Super Mario RPG's machine translation is filled with content that would make the video game ratings board cry.

Even though it uses plenty of helpful *kanji*, *Final Fantasy VI* still transforms into an amusingly bizarre mess when run through a machine translator.

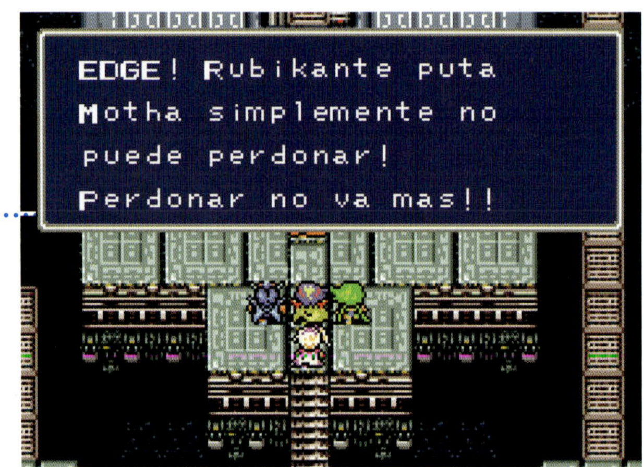

Japanese-to-English machine translations aren't the only game in town. Other language pairs like Japanese-to-Spanish and Korean-to-German can be just as entertaining!

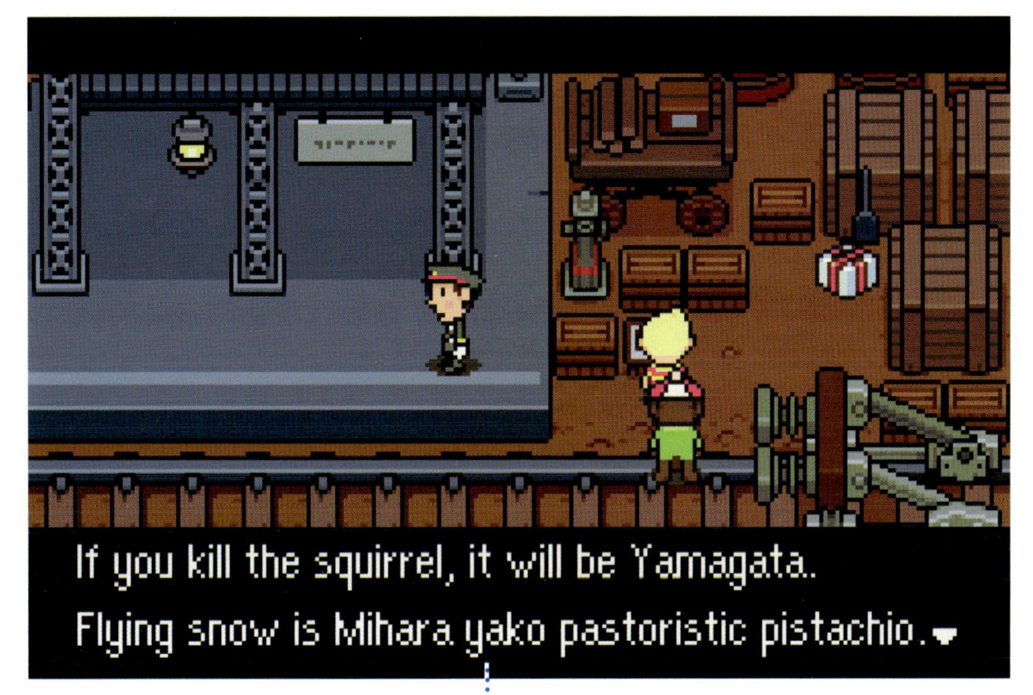

What happens when desperate fans, neglected games, and machine translations mix? I'm not sure there's an English word for it, so let's call it "sombatservation".

A Funky Future

The *Funky Fantasy IV* project has taught me a lot about a subject I never thought I'd study before. Machine translation technology will continue to transform, of course, so I'd love to keep an eye on how it grows. Maybe I'll keep running *Final Fantasy IV* through Google Translate from time to time to see how things change. And, who knows, if these rapid changes keep up, maybe I'll look at other machine translations in future books. There's so much to look forward to in the weird world of translation!

Check out these other funky *Legends of Localization* books!

Legends of Localization Book 1: The Legend of Zelda

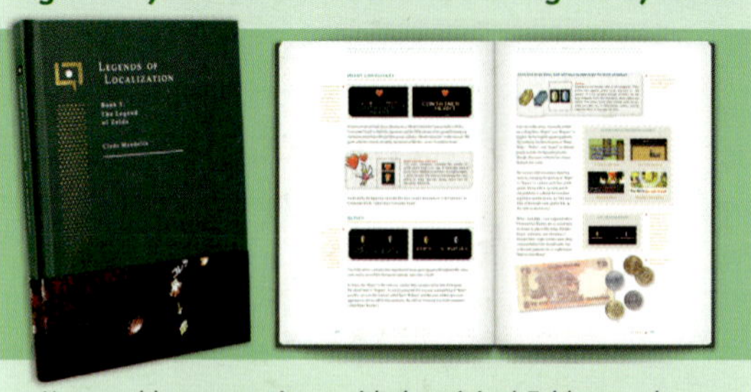

How would your experience with the original *Zelda* game have changed if you'd been born in Japan versus America or vice-versa?

208 pages • Bonus postcard included!

Legends of Localization Book 2: EarthBound

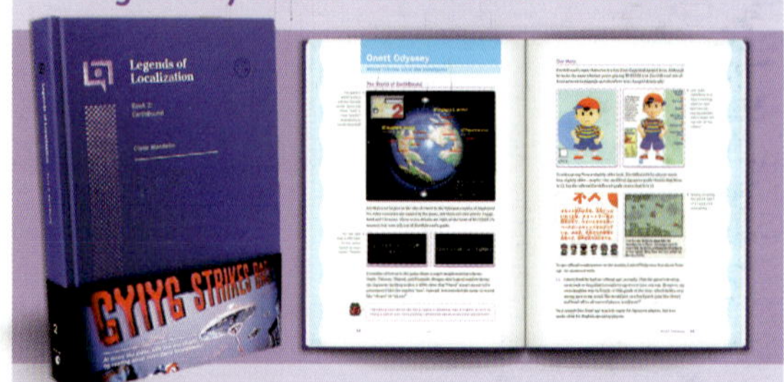

I started writing a book about *EarthBound*'s legendary translation but wound up creating something much bigger. You gotta read this!

432 pages • Scratch 'n sniff card included!

This be book bad translation, video games!

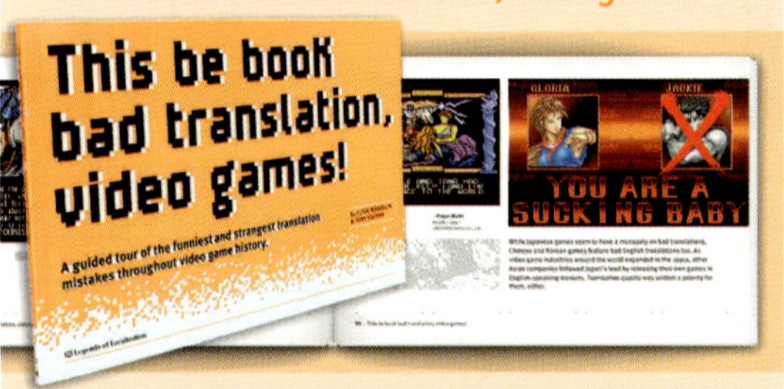

We've all heard of "All Your Base Are Belong to Us", but what is the history of bad translations, and how have they changed over time?

64 pages • 99 bad translations!

I'm Stuck in a Video Game

I had the "teacher attack chance" to document how a notable children's book got localized from Japanese into English!

Interviews • Concept art • Bonus stickers

Passport Series

The Legend of Zelda | *MOTHER 2*

If you're interested in learning Japanese but don't know where to start, these handy game-themed guides will teach you two writing systems, basic grammar, and more!